PC Audio Editing

PC Audio Editing

From broadcasting to home CD

Roger Derry M.I.B.S.

Focal Press

OXFORD AUCKLAND BOSTON JOHANNESBURG MELBOURNE NEW DELHI

Focal Press
An imprint of Butterworth-Heinemann
Linacre House, Jordan Hill, Oxford OX2 8DP
225 Wildwood Avenue, Woburn, MA 01801-2041
A division of Reed Educational and Professional Publishing Ltd

℞ A member of the Reed Elsevier plc group

First published 2000

© Reed Educational and Professional Publishing Ltd 2000

British Library Cataloguing in Publication Data
Derry, Roger
  PC audio editing
  1. Computer sound processing   2. Sound – Recording and
  reproducing – Digital techniques   3. Audiotapes – Editing
  I. Title
  006.5

Library of Congress Cataloguing in Publication Data
A catalogue record for this book is available from the Library of
Congress

ISBN 0 240 51596 X

Printed and bound in Great Britain
by Biddles Ltd, www.biddles.co.uk
Composition by Scribe Design, Gillingham, Kent

Contents

Foreword

Desktop audio, hard disk recording, non-linear recording ... no matter what you call it, recording, editing and producing audio art on the personal computer is here, with a big-time vengeance. Why? First off, the process is cost effective from the standpoint that edits, effects and mixes can often be done, undone and/or recalled from disk in a fraction of the time, when compared to analogue audio, not to mention that you don't have to spend an arm and a leg to get pro results. The second (and most important) reason is flexibility. Using a multitrack editor (or some special, fancy footwork on a 2-channel editor) you can import, record and/or edit all of the necessary tracks, add effects to any track that could use some spice, mix them down, and maybe even burn them to CD. Since the sound files themselves are often unaltered on the hard disk (a process known as non-destructive editing), edits and level changes can be made at any future time, by simply recalling the session from disk, making the changes and Voila! You're back in biz!

The bottom line is, those of you who are new to the process of PC-based hard disk recording are really in for a treat. This book will help you to gain a greater understanding of the various hardware, software and production aspects that are needed to make music, news and broadcast programmes, audio-for-visuals, you name it! It's A–Z approach to audio editing on the PC is an excellent first step towards diving into this new and fun technology. Before you know it, you'll be making, undoing and redoing edits in a fraction of the time it would have taken to get the razor blades out of their boxes.

Finally, before you delve into Roger's informative and easy-to-understand treatment of this important subject, I'd strongly urge you to get out the CD and load demo programs (like Cool Edit Pro) onto your PC. As you read along, it's important that you take the time to play with the various functions, controls and effects. Go ahead, get your feet wet! Create your own mix, add effects, edit up a storm, make mistakes (you'll probably be able to undo them) ... but, most of all, have fun learning!

David Miles Huber

Preface

For years PCs struggled hard to cope with the prodigious demands of handling audio, let alone doing it well, or at any speed. Nowadays 'entry-level' PCs can provide facilities that, in previous decades, professional audio technicians would have killed for. Indeed, there is no doubt that a contemporary six-year-old could happily edit audio on a PC. But, as always with most skills, the trick is NOT doing it like a six-year-old child.

This book is aimed at people who wish to make audio productions for issue as recordings, or for broadcast, using a Windows PC with material acquired using portable equipment, as well as in studios. There are already plenty of books on Music Technology; this one is aimed at people who want to use speech as well; the making of cassette sales promotions, radio programmes or, even, Son et Lumière! The book does not expect you to be a technician although, if you are, there will still be plenty to interest you, as the book includes production insights to help you use your PC for producing audio productions in the real world.

Even those who are experienced in editing and mixing with quarter inch tape can find the change to editing audio visually on a PC daunting. However, while this does bring a change of skills, most of the new skills are more easily learned.

Visual editing is done statically rather than having to use the arcane dynamic skills of 'scrub' editing, required by quarter inch tape. Hearing an edit while scrubbing the tape back and forth was a skill that took weeks, or even months, to perfect. Some were doomed never to be able to hear clearly. With the PC, making mistakes is far less of a problem; correcting errors is rapid, and safe, unlike unpicking bits of tape spliced with sticky tape.

The author's early career included being a specialist tape editor. He is now a convert to the use of a PC for editing. His career has included all the jobs that go to make a radio programme including production, reporting, management, studio and location technical operations (and fetching tea!). He can therefore offer advice and experience over the whole range of audio production.

This book is not a manual for a particular PC editor, although it is firmly based around the widely used Cool Edit Pro (version 1.2). It attempts to show the basic principles of the new technology. While different editors have their operational differences, those basic principles remain constant. Many aspects of Cool Edit Pro are described in detail. Audio is audio. Digits are digits. Different programs may implement any task differently, better or worse, but the principle remains the same.

1

Visual editing

Anyone who is used to editing audio using quarter inch tape may approach editing on a PC with trepidation. PCs allow you to edit visually rather than using your ears to 'scrub' the audio back and forth to hear the edit point. Some Digital Audio Workstations provide a simulation of scrub editing, either using special hardware in the form of a search wheel, or an option to use the computer's mouse. These are often unsatisfactory because of tiny, but noticeable, processing delays between the mouse movements and hearing the result.

However, editing on a PC is a different medium and techniques change. After 25 years of quarter inch tape editing, I found that I took to editing visually like the proverbial duck to water. I did, however, have to overcome an emotional resentment at what seemed like a 'de-skilling' of the task.

My analogy is with word processing. In the days of manual typewriters, using carbon paper to make copies, there was a high premium on accuracy. It was also important to be able to type each key with an even pressure, so that the document had a professional look. This was a skilled thing to do and was not learnt in a day.

When word processors were introduced, the rules changed. It was soon discovered that it was more efficient to type as fast as possible and to clean up typos afterwards. Printouts take no notice of the key pressures that the typist has used. This has meant that even a person with no keyboard skills can – given a great deal of time – produce a document with a professional appearance.

So it is with audio editing; with plenty of experience and good training to become a skilled tape editor it is possible, very quickly, to find the precise edit point, mark, cut and splice, to get a good edit all the time. It is a skill. It has to be learned.

The PC offers visual tools to aid your editing. It also has the potential to be much more accurate. With reel-to-reel quarter inch analogue tape at 15 ips, using 60° cuts, the effective best accuracy is 1/120th of a second. 90° cuts can improve this but, in practice, not by much. Most edits are made using much lower accuracy.

In contrast, the PC can offer 'sample-rate' accuracy. This means that you can edit down to the resolution of a single number (sample) in the digital data. Using the standard professional sampling rates of 32 kHz, 44.1 kHz or 48 kHz, this gives a possible accuracy of 1/32 000th, 1/44 100th or 1/48 000th of a second! As has already been observed, most edits don't need this accuracy, so it is as well that you do not have to work to this resolution!

When viewing the sound file, as a whole, the PC offers you a static graphical display of your audio with level represented by the width of a line. For stereo, there are two variable width lines, side by side, representing the left and right channels. You can choose at what resolution you want to look at the audio. The width of the screen can encompass several hours or one-thousandth of a second.

Illustrated (Figure 1.1) is a 4-second chunk of mono audio. The text of what is being said has been added to the illustration. PC audio editors are not (yet) clever enough to transcribe speech from your recording!

At this sort of resolution, you can easily see the rises and falls in level corresponding to individual syllables. Yet we can zoom in to even more detail. The usual way to do this is to select the area by dragging the mouse – just as you would to select text in a word processor. You then zoom to the selection. In this example using the word 'and' (Figure 1.2) is enough to show the individual vibrations (Figure 1.3). The 'D' sound at the end is now extremely easy to see.

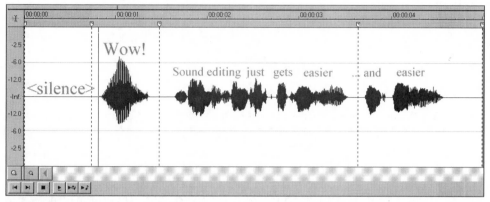

Figure 1.1

Figure 1.2

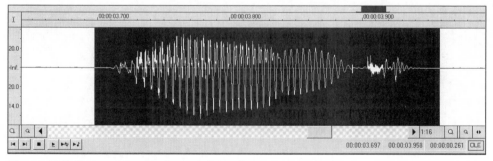

Figure 1.3 Zoomed in on the word 'and'

Cutting a section of audio is as easy as selecting it with a mouse and pressing the delete key, just as in word processing. Also, like word processing, you can cut, or copy, audio to a clipboard and then paste it elsewhere.

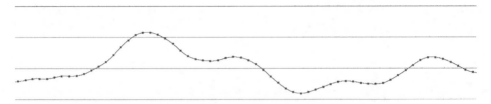

Figure 1.4 Audio zoomed in to showing samples as dots. The line joining them is created by the computer

There is usually an option for a 'mix' paste, where the copied audio is mixed on top of existing audio rather than being inserted. With several attempts and much use of the UNDO function, this can be a way of adding a simple music bed to speech but it lacks the sophisticated control available from using a non-linear editing mode giving you multitrack operation.

Most audio editors allow you to zoom to the sample level (Figure 1.4), but this is rarely of much value, except where you are manually taking out a single click from an LP transfer or some glitch picked up on the way. In general, noise reduction and declicking is more quickly done by software, although this will let through some clicks that have to be dealt with individually.

2

Some technical bits

There are some technical terms that are much bandied about in audio. While it is possible to survive without a knowledge of them, they are a great help in making the most of the medium. However, you may wish to skip this chapter and read it later.

2.1 Loudness, decibels and frequencies

How good is the human ear?

Sound is the result of pressure; compression/decompression waves travelling through the air. These pressure waves are caused by something vibrating. This may be obvious, like the skin of a drum, or the string and sounding board of a violin. However, wind instruments also vibrate. It may be blowing through a reed or by blowing across a hole; the turbulence causing the column of air within the pipe of the instrument to vibrate.

The two major properties that describe a sound are its frequency and its loudness.

Frequency
This is a count of how many times per second the air pressure of the sound wave cycles from high pressure, through low pressure and back to high pressure again (Figure 2.1). This used to be known as the number of 'cycles per second'. This has now been given a metric unit name: 1 hertz (abbreviation Hz) is one cycle per second and is named after Heinrich

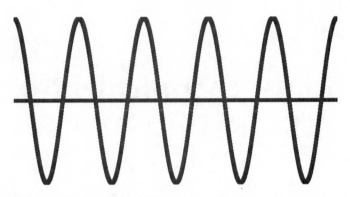

Figure 2.1 Five cycles of a pure audio sine wave

Hertz who did fundamental research into wave theory in the nineteenth century.

The lowest frequency the ear can handle is about 20 Hz. These low frequencies are more felt than heard. Some church organs have a 16 Hz stop which is added to other notes to give them depth. Low frequencies are hardest to reproduce and, in practice, most loudspeakers have a tough time reproducing much below 80 Hz.

At the other extreme, the ear can handle frequencies of up to 20 000 Hz, usually written as 20 kHz. (kilohertz). As we get older, our high frequency limit reduces. This can happen very rapidly if the ear is constantly exposed to high sound levels.

The standard specification for high fidelity audio equipment is that it should handle frequencies between 20 Hz and 20 kHz equally well. Stereo FM broadcasting is restricted to 15 kHz, as is the NICAM system used for television stereo in the UK. But Digital Radio is not, and so broadcasters have begun to increase the frequency range required when programmes are submitted.

The frequencies produced by musical instruments occupy the lower range of these frequencies. The standard tuning frequency 'middle A' is 440 Hz. The 'A' one octave below that is 220 Hz. An octave above is 880 Hz. In other words a difference of an octave is achieved by doubling or halving the frequency.

Instruments also produce 'harmonics'. These are frequencies that are multiples of the original 'fundamental' note. It is these frequencies, along with transients (how the note starts and finishes) that give an instrument its characteristic sound. This is often referred to as the timbre (pronounced 'tam ber').

Loudness

Our ears can handle a very wide range of levels. The power ratio between the quietest sound that we can just detect – in an impossibly quiet sound insulated room – and the loudest sound that causes us pain is:

1:1 000 000 000 000

'1' followed by twelve noughts is one million million! To be able to handle such large numbers a logarithmic system is used. The unit, called a bel, can be thought of as a measure of the number of noughts after the '1'. In other words, the ratio shown above could also be described as 12 bels. Similarly a ratio of 1:1000 would be 3 bels. A ratio of 1:1 – no change – is 0 bels. Decreases in level are described as negative. So a ratio of 1000:1 – a reduction in power of 1/1000th – is –2 bels.

In practice, the bel is too large a unit to be convenient. Instead the unit used every day, is one-tenth of a bel. The metric system term for one-tenth is 'deci', so the unit is called the 'decibel'. The abbreviation is 'dB' – little 'd' for 'deci' and big 'B' for 'bel' as it is based on a person's name – in this case Alexander Graham Bell, the inventor of the telephone and founder of Bell Telephones, who devised the unit for measuring telephone signals. Conveniently, a change of level of 1 dB is about the smallest change that the average person can hear (Table 2.1).

3 dB represents a doubling of power. 6 dB represents a doubling of voltage.

10 dB increases sounds twice as loud. 10 dB decreases sounds half as loud.

(Power equals Voltage × Current; double the voltage and you also double the current).

Table 2.1 Everyday sound levels

0 dB	Threshold of hearing: sound insulated room
10 dB	Very faint – a still night in the country
30 dB	Faint – public library, whisper, rustle of paper
50 dB	Moderate – quiet office, average house
70 dB	Loud – noisy office, transistor radio at full volume
90 dB	Very loud – busy street
110 dB	Deafening – pneumatic drill, thunder, gunfire
120 dB	Threshold of pain

2.2 Hearing safety

One of the hazards of editing on a PC is that it is often done on headphones in an office containing other people. Research has shown that people, on average, listen to headphones 6 dB louder than they would to loudspeakers. So already they are pumping four times more power into their ears.

When you are editing, there are going to be occasions when you wind up the volume to hear quiet passages and then forget to restore it when going on to a loud section. The resulting level into your ears is going to be way above that which is safe.

Some broadcasting organizations insist that their staff use headphones with built-in limiters to prevent hearing damage. By UK law, a sound level of 85 dBA is defined as the first action level (the 'A' indicates a common way of measuring sound-in-air level, as opposed to electrical, audio, decibels). You should not be exposed to sound at, or above, this level for more than 8 hours a day. If you are, as well as taking other measures, the employer must offer you hearing protection.

90 dBA is the second action level and at this noise level or higher, ear protection must be worn and the employer must ensure that adequate training is provided and that measures are taken to reduce noise levels, as far as is reasonably practicable.

The irony here is that the headphones could be acting as hearing protectors for sound from outside, but be themselves generating audio levels above health and safety limits.

If you ever experience 'ringing in the ears' or are temporarily deafened by a loud noise, then you have permanently damaged your hearing. This damage will usually be very slight each time, but accumulates over months and years; the louder the sound the more damage is done.

Slowly entering a world of silence may seem not so terrible but, if your job involves audio, you will lose that job. Deafness cuts you off from people; it is often mistaken for stupidity. Worse, hearing damage does not necessarily create a silent world for the victim. Suicides have been caused by the other result of hearing damage. The gentle-sounding word used by the medical profession is 'tinnitus' This conceals the horror of living with loud, throbbing sounds created within your ear. They can seem so loud that they make sleep difficult. Some people end up in a no-win situation where, to sleep, they have to listen to music on headphones at high level to drown the tinnitus. Of course this, in turn, causes more hearing damage.

If you are reading this book then you value your ears;

so please take care of them!

2.3 Analogue and digital audio

Analogue audio signals consist of the variation of sound pressure level with time being mimicked by the analogous change in strength of an electrical voltage, a magnetic field, the deviation of a groove, etc.

The principle of digital audio is very simple and that is to represent the sound pressure level variation by a stream of numbers. These numbers are represented by pulses. The major advantage of using pulses is that the system merely has to distinguish between pulse and no-pulse states. Any noise will be ignored provided it is not sufficient to prevent that

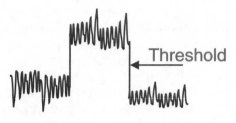

Figure 2.2

distinction (Figure 2.2). The actual information is usually sent by using the length of the pulse; it is either short or long. This means that the signal has plenty of leeway in both the amplitude of the signal (how tall) and the length of the pulse. Well-designed digital signals are very robust and can traverse quite hostile environments without degradation.

However, this can be a disadvantage for, say, a broadcaster with a 'live' circuit, as there can be little or no warning of a deteriorating signal. Typically the quality remains audibly fine, until there are a couple of splats, or mutes, then silence. Analogue has the advantage that you can hear a problem developing and make arrangements for a replacement before it becomes unusable.

Wow and flutter are eliminated from digital recording systems, along with analogue artefacts such as frequency response and level changes. If the audio is copied as digital data then it is a simple matter of copying numbers and the recording may be 'cloned' many times without any degradation. (This applies to pure digital encoding; however, many modern systems, such as Minidisc and Digital Radio, use a 'lossy' form of encoding. This throws away data in a way that the ear will not usually notice. However, multi-generation copies can deteriorate to unusability within six generations. Even CD and DAT are lossy to some extent, as they tolerate and conceal digital errors.)

What is digital audio?

Pulse and digital systems are well established and can be considered to have started in Victorian times.

Perhaps the best known pulse system is the Morse code. Like much digital audio, this code uses short and long pulses. In Morse these are combined with short, medium and long spaces between pulses to convey the information. The original intention was to use mechanical devices to decode the signal. In practice it was found that human operators could decode by ear faster.

Modern digital systems run far too fast for human decoding and adopt simple techniques that can be dealt with by microprocessors with rather less intelligence than a telegraph operator. Most systems use two states that can be thought of as on or off, short or long, dot or dash (some digital circuits and broadcast systems can use three, or even four, states but these systems are beyond the scope of this book).

The numbers that convey the instantaneous value of a digital audio signal are conveyed by groups of pulses (bits) formed into a digital 'word'. The number of these pulses in the word sets the number of discrete levels that can be coded.

The word length becomes a measure of the resolution of the system and, with digital audio, the fidelity of the reproduction. Compact disc uses 16-bit words giving 65 536 states. NICAM stereo, as used by UK television, uses just 10 bits but technical trickery gives a performance similar to 14 bit. Audio files used on the Internet are often only 8 bit in resolution while systems used for telephone answering, etc. may only be 4-bit systems or less (Table 2.2).

Table 2.2 Table showing how many discrete levels can be handled by different digital audio resolutions

1 bit=2	7 bit=128	13 bit=9192	19 bit=524 287
2 bit=4	8 bit=256	14 bit=16 384	20 bit=1 048 575
3 bit=8	9 bit=512	15 bit=32 768	21 bit=2 097 151
4 bit=16	10 bit=1024	16 bit=65 536	22 bit=4 194 303
5 bit=32	11 bit=2048	17 bit=131 071	23 bit=8 388 607
6 bit=64	12 bit=4096	18 bit=262 143	24 bit=16 777 215

The larger the number, the more space is taken up on a computer hard disk or the longer a file takes to copy from disk to disk or through a modem. Common resolutions are: 8 bit: Internet, 10 bit: NICAM, 16 bit: compact disc, 18/20 bit: enhanced audio systems for production.

Sampling rate

A major design parameter of a digital system is how often the analogue quantity needs to be measured to give accurate results. The changing quantity is sampled and measured at a defined rate.

If the 'sampling rate' is too slow then significant events may be missed. Audio should be sampled at a rate that is at least twice the highest frequency that needs to be produced. This is so that, at the very least, one number can describe the positive transition and the other a negative transition of a single cycle of audio. This is called the Nyquist limit after Harry Nyquist of Bell Telephone Systems who developed the theory.

For practical purposes, a 10 per cent margin should be allowed. This means that the sampling rate figure should be 2.2 times the highest frequency. 20 kHz is regarded as the highest frequency that most people can hear. This led to the CD being given a sampling rate of 44 100 samples a second (44.1 kHz – the odd 100 samples per second is a technical kludge. Originally digital signals could only be recorded on video-type machines. 44.1 kHz will fit into either American or European television formats).

It is vital that the sampling process is protected from out-of-range frequencies. These 'beat' with the sampling frequency and produce spurious frequencies that not only represent distortion but, because of their non-musical relationship to the intended signal, represent a particularly nasty form of distortion. These extra frequencies are called alias frequencies and the filters called anti-aliasing filters. It is the design of these filters that can make the greatest difference between the perceived quality of analogue to digital converters. You will sometimes see references to 'over-sampling'. This technique emulates a faster sampling rate (4×, 8×, etc.) and simplifies the design of the filters.

Errors

A practical digital audio system has to cope with the introduction of errors, owing to noise and mechanical imperfections, of recording and transmission.

These can cause distortion, clicks, bangs and dropouts, when the wrong number is received.

Digital systems incorporate extra 'redundant' bits. This redundancy is used to provide extra information to allow the system to detect, conceal or even correct the errors. Decoding software is able to apply arithmetic to the data, in real time, so it is possible to use coding systems that can actually detect errors.

However, audio is not like accountancy, and the occasional error can be accepted, provided it is not audible in a well-designed system. This allows a simpler system, using less redundant bits, to be used to increase the capacity, and hence the recording length, of the recording medium. This is why computer CD-ROMs storing computer programs have a lower capacity than CD-audio. CD-ROMs have to use a more robust error correction system as NO errors can be allowed. As a result a standard CDR can record 720 Mbytes of CD audio (74 minutes) but only 650 Mbytes of computer data.

Having detected the error a CD player may be able to:

1 Correct the error (using the extra 'redundant' information in the signal).
2 Conceal the error. This is usually done either by sending the last correctly received sample (replacement), or by interpolation, where an intermediate value is calculated.
3 Mute the error. A mute is usually preferable to a click.

While the second and third options are good enough for the end product used by the consumer, you need to avoid the build-up of errors during the production of the recording. Copies made on hard disk, internally within the computer, will be error-free. Similarly copies made to CD-ROM or to removable hard disk cartridge, will also have no errors (unless the disk fails altogether because of damage).

Multi-generation copies made through the analogue sockets of your sound card will lose quality. Copies made digitally to DAT will be better, but still accumulate errors. However, computer backup as data to DAT (4 mm) should be error-free (as should any other form of computer backup medium. These have to be good enough for accountants, and therefore error-free).

How can a computer correct errors?
It can be quite puzzling that computers can get things wrong but then correct them. How is this possible? The first thing to realize is that each datum bit can only be '0' or '1'. Therefore, if you can identify that a particular bit is wrong, then you know the correct value. If '0' is wrong then '1' is right. If '1' is wrong then '0' is right.

The whole subject of error correction involves deep mathematics but it is possible to give an insight into the fundamentals of how it works. The major weapon is a concept called parity. The basic idea is very simple, but suitably used can become very powerful. At its simplest this consists of adding an extra bit to each data word. This bit signals whether the number of '1's in the binary data is odd or even. Both odd and even parity conventions are used. With an even parity convention the parity bit is set so that the number of '1's is an even number (zero is an even number). With odd parity the extra bit is set to make the number of '1's always odd.

Received data is checked during decoding. If the signal is encoded with odd parity, and arrives at the decoder with even parity, then it is assumed that the signal has been corrupted. A single parity bit can only detect an odd number of errors.

Quite complicated parity schemes can be arranged to allow identification of which bit is in error and for correction to be applied automatically. Remember that the parity bit itself can be affected by noise.

Illustrated in Figure 2.3 is a method of parity checking a 16-bit word by using eight extra parity bits. The data bits are shown as 'b', the parity bits are shown as 'P'.

The parity is assessed both 'vertically' and 'horizontally'. The data are sent in the normal way with the data and parity bits intermingled. This is called a

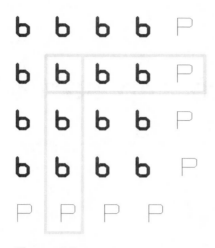

Figure 2.3

Hamming code. If the bit in column 2 row 2 were in error, its two associated parity bits would indicate this. As there are only two states then, if the bit is shown to be in error, reversing its state must correct the error.

Dither

The granular nature of digital audio can become very obvious on low level sounds such as the die-away of reverberation or piano notes. This is because there are very few numbers available to describe the sound, and so the steps between levels are relatively larger, as a proportion of the signal.

This granularity can be removed by adding random noise, similar to hiss, to the signal. The level of the noise is set to correspond to the 'bottom bit' of the digital word. The frequency response of the hiss is often tailored to optimize the result, giving noise levels much lower than would be otherwise expected. This is called 'noise-shaping'. PC audio editors often have an option to turn this off. Don't do this unless you know what you are doing. Dither has the, almost magical, ability to enable a digital signal to carry sounds that are quieter than the equivalent of just 1 bit.

3

Hardware and software requirements

3.1 PC

Audio editing on a PC became a practical proposition for professional use once the equivalent or better than a Pentium II running at more than 200 MHz was reached. Even a relatively slow machine like this, will give more than adequate speed of processing and is able to perform multitrack mixing.

Consider getting a 'full tower' case as this will then conveniently sit on the floor releasing space on your desk. Floor standing towers can also more easily be strapped down if theft is likely to be an issue.

3.2 Sound card

This should be capable of at least 16-bit audio at sampling rates of 44.1 kHz (CD standard) and 48 kHz (DAT standard). Lower rates will be useful if you need to audition files intended for the Internet as well as for reproducing noises made by other programs and by Windows operations.

If you have access to high quality analogue sources such as 15/30 ips Dolby SR recordings, or high quality microphones in quiet studios, then a higher bit rate is highly desirable; even if your finished recording is going to be 16 bit. This is because headroom will have to be allowed at the time of recording to prevent digital overload. Quite likely a few levels will have been increased during post production. This means that you can end up delivering the equivalent of a 12-bit recording! (See section on analogue transfer on page 52.)

As well as conventional analogue inputs and outputs, it can be useful if your sound card will handle digital inputs and outputs so that your original digital recordings can be 'cloned' on to your computer and your final mix returned without quality loss.

In practice, many of the domestic machines used by many professionals have only optical digital outputs rather than the electrical outputs required by most sound cards. However, you can buy converter boxes which will do the job very adequately but at the expense of yet another box cluttering up your desk.

Optical connectors use modulated red light rather than electricity to transfer the data. This light is visible to the naked eye when a connector is carrying an output.

Be aware that there two physically different optical standards. One is called TOSLINK and the other is the same shape and size as minijack audio connectors. The socket is usually dual function and can be used for electrical analogue audio or for optical digital (see page 48).

Sound cards are traditionally slotted into the back of the computer, after removing the cover. This is probably as good a place as any; the card is kept safe and not likely to be dropped. However, various boxes that plug into the USB bus of the computer can be used instead. These have the advantage that they can be plugged and unplugged without switching off the computer. While some versions of Windows 95 claim to be able to use the USB bus, you really need Windows 98 or later for success.

3.3 Loudspeakers/headphones

Ordinary PC 'games' loudspeakers are not adequate to assess sound quality. You should pay the extra for music quality active speakers ('active' means that the speakers have their amplifiers built in for driving directly from the sound card output). If you have the space this can be a proper hi-fi system with the computer's sound card feed a line level 'aux' input.

Much of the time, you may be operating in a shared office using headphones. These should be of the best quality with a decent bass response. Try always to check mixes on loudspeakers, if at all possible, as headphones give a very different impression to loudspeakers. As a generalization, a mix that sounds good on loudspeakers will sound good on headphones. The reverse is not always true.

Headphones are much more critical of poor editing than loudspeakers. The standard advice to broadcasters used to be, not to worry if an edit was audible on headphones but not on loudspeakers. However, with the popularity of Walkman-style cassette and Minidisc players with radios, it is no longer possible to assume that all your audience is listening on loudspeakers.

An exception to this is when the audience is known to be listening on headphones – walk-round cassette guides to exhibitions, for example. Here headphones will give you accurate results. Use your best headphones for production, but do check what the mix sounds like on the, probably cheaper, headphones used at the show.

3.4 Hard disks

Ideally you should have a second separate hard drive for your audio data. Even better, for the fastest operation, yet another separate disk for handling the audio editor's temporary files. These drives should be of audio-visual (AV) quality especially if you intend to 'burn' compact discs. Early, non-AV quality disks recalibrated themselves at inconvenient times, to compensate for temperature changes, interrupting a continuous flow of audio (or video). In practice, it is rare for modern hard disks not to be suitable for AV but you should confirm with your supplier that the drives supplied are suitable for the task.

Modern IDE/EIDE drives are adequate for audio purposes although many people still think that the SCSI system is well worth the extra cost for AV work. SCSI also has the advantage of being able to handle up 15 devices, which can include scanners as well as hard drives, rather than the four IDE drives that most computers support.

If you are likely to be working on several projects at once, then a removable hard disc cartridge system will be useful. 1 Gbyte or greater cartridges can hold enough audio to make a CD – including spare material and auxiliary files. These are, however, more expensive than the commonly available 100 and 250 Mbyte cartridges. However, those lower capacity media are able to take 10 or 25 minutes of CD quality stereo audio, which is entirely adequate for short items.

While less convenient, another alternative is to use standard hard drives mounted in removable caddies. This way, tens of gigabytes of data can be carried from computer to computer, provided that the same caddy system is used in each computer. The cheapest of these require you to turn off the computer to change caddies, but systems are available that allow 'hot' swapping and will update the Windows drive list on-the-fly. The caddies are more expensive than they ought to be, as most people will need more drive carriers than slots to put them in, but the two halves of the system are difficult to buy separately.

3.5 Audio editor

The audio editor is the key to the whole operation. There are any number of different ones made by a host of manufacturers. These range from full-scale Digital Audio Workstations (DAWs), often using proprietary hardware, to very simple 'freebies' with the minimum of facilities.

Most of the examples in this book will be taken from a good example of presently available software. Cool Edit Pro is a combined linear/non-linear editor potentially able to handle 64 audio tracks. It is a good all-round general workhorse. Its methods are easily transferred to other editors.

3.6 Linear editors

Editors fall into two methods of working: linear and non-linear. Linear editors can handle one stereo, or mono, audio track at a time. They operate like word processors; any editing is destructive, a deleted section is actually deleted from the audio file (although, as with word processors, there is likely to be an 'undo' feature available). When saved, any cut material is lost, unless a backup copy of the original has been kept.

Because of the large amounts of data involved with audio, cut and paste are instantaneous only for the smallest of sections. Copying a 40 minute stereo file can take several minutes. While the process may be much quicker than with pre-digital technology, minutes spent staring at a progress bar crossing your screen are frustrating and unproductive (see 'Blue bar blues' on page 76).

3.7 Non-linear editors

Non-linear editors do not change the audio files being edited in any way. Instead they create 'Edit Decision Lists' (EDLs). They don't play an audio file linearly from beginning to end. Instead the files are played out of sequence – non-linearly – with edits performed by skipping instantly to the next section.

The most obvious advantage is that material is never lost. But this can be a disadvantage as well, because your hard disk can soon become cluttered with unwanted material. In practice, some culling with a linear editor is useful before using a non-linear one.

You can have any number of different EDLs for the same audio, thus you can have any number of differ-ent versions. You can decide to keep a version that you are reasonably happy with, but continue to see if you can further improve the item. This is ideal if the same item has to be repackaged for different programme outlets. A programme trail can be one edit decision list and the broadcast programme another.

While this is possible with linear editors, it is at the expense of much duplication of audio data, which can soon fill a hard disk. Only the non-linear editor can give you the opportunity to 'unpick' edits for your new version.

Non-linear editors can also handle more than one audio file at a time and play from different sections of a single audio file simultaneously. In this way cross-fades and mixes can be performed, without permanently changing the original audio material.

Non-linear editing operations, in general, can be faster as the computer is not having to move multi-megabyte chunks of digital audio data around the hard disk. Instead, the changes in the EDL are measured in just tens of bytes.

3.8 Multitrack

You don't have to handle complex EDLs yourself as the non-linear editor presents itself, on screen, as if it were a multitrack tape recorder. The lists are managed by the editing software not by you. However, it is far more flexible than a physical multi-track tape recorder.

The non-linear editor shares, with multitrack, the ability separately to record individual tracks in synchronization with existing ones. You can also 'drop-in' to replace a section. But, unlike hardware multitrack with its physical restraints, you can also 'slide' tracks back and forth with respect to each other. Track 'bouncing' (copying from one track to another) is virtually instantaneous. You can add any level changes and crossfades in a way that would need full-scale automation in a studio using physical multitrack.

3.9 Audio processing

The standard tricks of an audio editor are cut and paste, as in a word processor, and the ability to cross-fade between two pieces of audio.

In addition, audio editors for professional and semi-professional use come with a barrage of special effects. Some you will have daily use for, others may 'get you out of a jam' some time. Many editor programs have the facility for their user to buy 'plug-ins' that will add specialist, or enhanced quality effects.

The most important processing options are:

- *Normalization*. This standardizes the sound level of each item, although not necessarily the loudness.
- *Reverberation*. Sometimes known as 'artificial echo', this will add a room, or hall, acoustic to your recording. This is most often used with music recording. With speech programmes you are less likely to need it, except to get you out of a jam by recreating an acoustic. There is also the annual ritual of adding reverberation to witches' voices for Halloween.
- *Compression*. This reduces the range of volume between the quietest and loudest sound. Used with care, it can enhance your recording. Used carelessly it can make your item offensively difficult to hear. The overall effect is to make your material louder while remaining within the constraints of the maximum digital level available.
- *Noise reduction*. This should not be confused with the noise reduction used on analogue tape systems like dbx, Dolby A, B, C, S or SR. These are 'companding' systems where programme material is compressed for recording but expanded back to its original dynamic range on playback. (Dolby A and Dolby SR are used in professional studios. Dolby B, C and S are domestic systems.) A digital system does not need such techniques. However, you may have a digital recording copied from a poor original, either from the archive, or one

afflicted with technical problems at the time of the original recording. While intended to remove such recording problems as tape hiss, your editor's noise reduction system may well turn out to be useful in removing acoustic noise such as traffic rumble or air conditioning whine.

- *Declicking.* Old 78 rpm recordings and vinyl LPs are plagued by clicks as well as noise and will benefit from electronic declicking. However, don't be too enthusiastic about cleaning up old recordings as they often need to sound old in the context of a programme item.
- *Filters.* These are useful for filtering out constant sounds as well as sound at the extreme ends of the frequency range. Cutting the bass will remove rumbles coming through the wall, or from underground trains. Mains hum can be a problem and special 'notch' filters are available to remove the worst of this. Different settings are needed for American (60 Hz) and European (50 Hz) mains systems.
- *Equalization.* This is a sophisticated form of 'tone' control usually referred to as 'EQ'. This can be used to brighten up muffled recordings, or to reduce sibilance from some interviewees. Domestic tone controls normally only affect the top and bottom frequencies; equalizers can also do useful things to the middle frequencies.

3.10 Mastering

Mastering refers to the process of creating a recording that is usable away from the PC. It is possible to play, or broadcast, your mix from the computer directly to an audience. However, this does tie up your computer, with all its expensive resources.

You need to put your final mix onto a 'removable medium' in a conventional stereo form. This also has the, not inconsiderable, advantage that no one else can 'muck about with' your item.

Mastering can be as simple as playing the mix out of the computer's sound card onto a normal domestic medium like compact cassette. This format has the major advantage of being playable almost universally as the compact cassette is the most popular sound recording medium yet devised.

Professionals may copy to DAT (preferably using a digital link) or to quarter inch tape. Both these operations have to be done in 'real time' – 30 minute programmes take half an hour to dub.

In a broadcast environment, even now, dubbing to quarter inch has the advantage that, if an edit has been missed, or a retake forgotten, then a quick, old-fashioned razor blade cut can be made. This is quicker than correcting the edit on the computer and then having to start copying to DAT from the beginning.

3.11 CD recording software

Compact disc has become so widespread that it becomes a logical medium for distributing your item. Commercial CDs are pressed, not recorded individually and, for a long time, this route was not practical.

The advent of the recordable CD (CDR) has therefore been a boon to all creators of audio material. However, it is still constrained by a number of factors.

Some early ordinary audio CD players may not recognize a CDR. This, in fact, is relatively unusual and is unlikely to be a problem in a professional environment.

There are many different types of CD: CD-ROM, multi-session CD, etc. Domestic CD audio players (and many professional ones as well) only recognize one CD format, the original audio format. A consequence of this is that any compact disc that is to be used on a standard audio player needs to be recorded in a tightly defined way (as the process involves a laser making physical changes to the disc, it is usual to talk of 'burning' a CD).

CD burning is not something that can be interrupted and this means that many computers cannot be used for anything else while the CD is being made. However, the good news is that the CD can be burnt in less than real time. Computer CD recorder drives that can handle ×2, ×4, ×6 and ×8 are common. In itself, a fast drive is not enough; the computer system itself must be fast enough to be able to provide the data without a break. If it fails, even momentarily, then the whole burn fails and the disc is useless.

Failed CDRs are often referred to as 'coasters' or 'table mats', as all they are good for is putting under cups of coffee. However, recorder speeds of up to ×4 should be well within the capabilities of all but the most ancient of machines.

There are also stand-alone CD recorders that take an audio or digital audio signal and record discs in real time. Many of these will only use so-called 'copyright paid' consumer blanks which cost up to six or eight times that of computer data blanks.

You can also buy 'rewritable' CDs that can be erased when the data they contain is finished with. They can be recorded as standard audio CDs but, while most computer CD-ROM drives can play them, very many audio CD players cannot.

Most CD writers come bundled with basic software. This is usually a 'sawn-off' version of generally available software with some of its cleverer options removed, so that you will be encouraged to go out and buy the full edition.

Simple software often gives you two main options for burning a complete CD. These are Track At Once (TAO) and Disc At Once (DAO). Track At Once means literally this: each CD track is burnt as a separate action by the software, with the laser being turned off between tracks. This usually imposes 2-second pauses between tracks. Disc At Once burns the entire disc in one go without switching off the laser. With simpler software this provides no gaps at all between tracks.

3.12 DVD

At the time of writing, DVD has still to fulfil its potential although it promises to have most of the advantages of CD but with five to ten times the capacity. However, it will be some time before playback machines are as widespread as CD players are at present.

3.13 MIDI

Many digital audio editors are also set up to handle MIDI. If you are only using audio, then these facilities can be ignored. MIDI is only useful if you are planning to include synthesized music sessions within your programme, or you are synchronizing the audio to other events, such as a Son et Lumière.

The term MIDI stands for Musical Instrument Digital Interface. Do not confuse it with the term midi as applied to hi-fi units, which is a description of their size (midi as opposed to 'biggi').

The connections are via DIN plugs which, although physically identical to audio DIN plugs, are differently wired. A typical synthesizer will have three sockets as shown in Figure 3.1. The 'IN' will drive the synthesizer (sound making) circuits. The 'OUT' will carry the output of the keyboard.

The MIDI system allows up to 16 instruments to be connected together in a 'daisy chain' (Figure 3.2). The 'THRU' socket (American spelling of 'through') relays the data arriving at the IN socket, unchanged, onto the next instrument. In the process the electrical signal is 'cleaned up' and isolated so that a fault on one machine will not necessarily adversely affect the rest of the instruments in the chain.

MIDI data records performance data NOT audio. It is the modern electronic equivalent to the punched

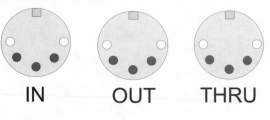

IN   OUT   THRU

Figure 3.1

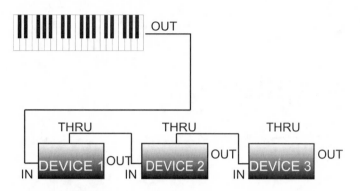

Figure 3.2

sheets that control player pianos and fairground organs.

MIDI can be used to synchronize different PC programs together. For specialized purposes, like exhibitions and Son et Lumière, MIDI can be used for external control of the audio output.

4

Acquisition

4.1 Responsibilities

All the equipment, bells, whistles and toys are of no use whatsoever unless there is something worthwhile to record, something your listeners will want to hear. Acquiring that material is the first task.

At the more complex level, you may be editing something that has been put together in a studio, or maybe a recording of a live event. In the context of this book, this means that acquisition is easy; you are presented with a stereo mix on a DAT or even a multitrack mix on an eight-track video cassette-based digital multitrack.

Your first responsibility is to ensure that you can handle the delivery format. An eight-track ADAT tape is fine only if you have an ADAT machine and your computer is set up to record its tracks, either by analogue or using the optical ADAT eight-track digital interface.

Even with DAT, you should have specified the sampling rate you require. This should normally be the rate you intend to use for the final edited version: 44.1 kHz for compact disc related work or, maybe, 48 kHz for material destined to be broadcast. Do not use lower sampling rates, or bit rates lower than 16, even if material is destined as a reduced data Internet sound file. You never know when there will be another use for the audio and the better the input to a poor reproduction system the better it will sound.

However, much of the material put together with a PC audio editor is likely to be less ambitious, and self-acquired; probably using a portable recorder, preferably DAT or Minidisc, to allow digital transfer to your computer.

A common way of obtaining your material is the interview; so here are a few guidelines to get the best out of them. These guidelines are as applicable to the promotional sales interview with the chairman of the company, as for a news interview for local or national radio.

4.2 Interviewing people

Preparation

Editing starts here, so consider the following; so as not to waste your own or other people's time and resources:

What is the purpose of the interview?

Are you interviewing the right person? (Do they have the information you want? Do they have a reasonable speaking voice?) Remember that the boss's deputy or assistant often has a better finger on the pulse of day-to-day problems so it is often useful to interview both, if you can.

Prepare your subject area for questions,

rather than a long list of specific questions.

Choose a suitable, and convenient, location.

A senior manager's office is usually suitable, if it has a carpet and soft furnishings; but it is extremely important to get the person out from behind the desk. While you will tell them that acoustics is the reason, equally vital is that people speak differently across desks. Sitting beside them on a sofa will usually get a much more human response.

Type of interviews

- *Hard* – to expose reasoning and to let the listener make up his/her own mind. Interviewee comes in cold, with no knowledge of questions. Commonly used with politicians or those in the public eye.

- *Informational* – aim is to get as much information as possible, so this is likely to be a more friendly, conversational style of interview. You may prepare your interviewee with a warm-up by outlining the areas of questioning.
- *Personality* – here you are interested in revealing the personality of the interviewee.
- *Emotional* – probably the most difficult type of interview requiring the greatest tact and diplomacy from the interviewer. Such interviews are of the 'How do you feel?' variety used when interviewees may be under great stress following a tragedy, e.g. Hillsborough, Clapham, Lockerbie, Soho, Paddington (although try to avoid the actual question 'How do you feel ...?')

Technique

Make sure that you actually know how to use the recorder! Try it out before the day of the interview. Test the recorder before you leave base. Test it when you arrive and are waiting for your interviewee. Test it when you take level. Always keep the test recordings.

Always record useful information such as the date, who you are going to interview. This will help speed the identification of a recording when the label has fallen off and when it has been transferred to computer (some DAT and Minidisc machines will record a time of day and date code on the recording. This can be useful, provided you remembered to set the clock when you took the machine out of its box).

NEVER say anything rude about the person or company. 'I'm off to interview that prat Bloggs of that useless Bloggo company' might get you a titter in the office, but loses its edge when you find yourself accidentally playing it back to Mr Bloggs after he has given you level. Machines that only offer headphone monitoring are a help here but spill from headphones can be very audible.

If using a handheld microphone, then sit, or stand, close to interviewee. Get them out from behind their desk! The microphone should be 9 to 12 inches from interviewee and a similar distance from you. If you have to compromise, favour the interviewee.

Listen for, and anticipate, external noises – e.g. children playing, dog barking, phone ringing, interruptions.

Relax the interviewee, if necessary, by discussing areas of questioning.

Keep eye contact and make appropriate silent responses such as nodding or smiling.

Ask short, clear questions, one at a time. Remember we want to hear the interviewee not the interviewer.

Listen carefully and keep your questions relevant to what is being said. Ask open questions – who, what, where, why, how, is it on the increase?

Pre-recorded interviews

Record background atmosphere for 15–30 seconds before and after interview to use when editing. In practice, leaving a 2-second gap between 'Right we're recording' and the first question will give you the cleanest pause.

If you are using two microphones with a stereo recorder, then deliberately record the interview 'two-track' so that one voice is on one track and yours on the other. Keep to a convention so that questions are always on the same track. 'The presenter is always right' was the tongue-in-cheek convention used in BBC Radio. In the days of quarter inch tape this also had the advantage that the answers could be heard on an old half track mono machine which only read the top (left-hand channel) of the tape.

Record the interviewee's name and title. Preferably get them to do it themselves and use this as a way of setting the level on the recorder. As well as giving you a factual check – are they Assistant Manager or Deputy Manager – it gives you a definitive pronunciation of their name; is Mr Smyth pronounced Sm-ith or Sm-eye-th? In some styles of documentary this can be used for them to introduce themselves rather than this being done by the presenter.

After the interview play back the last few seconds to ensure that the piece has recorded OK. Ask the interviewee if there is anything that they feel was left out.

In the trade, interviews done with people randomly selected in the street are known as vox pops (from the Latin *vox populi;* voice of the people). They present their own unique problems.

Vox pop interviews

Always test the equipment before leaving base.

Follow a closed question (e.g. 'Do you think...?') with an open question (e.g. 'Why do you think that?) to get interviewees to justify their opinions.

Prepare several forms of the same question to put to interviewees.

Choose a suitable location away from traffic, road works, etc. (unless, perhaps, the item is about the traffic or chaos caused by the road works).

Approach interviewees with a brief explanation of why you're conducting a survey of public opinion.

Keep the recording level the same and adjust the microphone position for quieter or louder interviewees.

Record background actuality at beginning and end of tape.

(A comprehensive guide to interviewing is contained in *Radio Production – a manual for broadcasters,* fourth edition, by Robert McLeish, Focal Press, Oxford, 1999.)

4.3 Documentary

Documentary material will also use 'actuality'. This is material recorded, on location, of activity which can be used both for illustration and for bridging passages of time.

Record lots of it! It is too easy to be so closely focused on the interview material that the non-vocal material is forgotten. Do not be in the situation of the producer who recorded a feature about a mountain climbing expedition and forgot to record any sounds of people climbing! (In this particular case the situation was saved by a single 2-second sequence being looped, repeated and used in a stylized way.)

Actuality is different from sound effects. Actuality is the real thing, recorded at the actual location. Sound effects come from a library and are used to fake actuality. This is fine for drama but has only limited use in factual items. Using an FX disc of a Rolls Royce driving off would be fine as an illustration of a car driven by rich people, but not ethical if cued as 'Joe Bloggs driving off in his Rolls Royce after my interview'.

Music can enhance an item, but beware of copyright and performance rights. While the sound of a busker playing in the background is unlikely to be a problem, using a complete performance of a piece of music usually is.

4.4 Interviews on the move

Perhaps not so much an interview as two or more people reacting to an environment. This could be observing badgers at night, a historical house or a carnival. The object is to capture as much of the atmosphere as possible while not making it totally impossible to edit.

An ideal situation would be a four-track recording. Two tracks would be fed from a stereo microphone to record atmosphere; the other two would be fed by two mono microphones, one for the presenter – possibly a personal 'tie-clip' mic – and the other handheld for whoever else is talking from moment to moment. An alternative might be two small port- ables – Minidisc or DAT – one recording atmosphere and the other recording dialogue.

The disadvantage of such an arrangement is that the technology is in danger of taking over. The person holding the mics and carrying the recorders may be being used for their knowledge of the subject rather than their technical skill. Here, a separate recorder operated by the producer can help out with collecting actuality.

However, it is very possible to get a dramatic, and memorable, recording using a single handheld stereo mic but it is vital that the mic is not much moved about with respect to the speakers as, otherwise, their voices will go careering around the sound stage. In noisy environments, the mic should be as close as feasible to the voices in order to get as much separation as possible so that what is being said can be understood. Plenty of actuality should be obtained for dubbing over the edited speech. It is easy to add sound, impossible to take away. A sensible discipline is also required so that the unrepeatable sounds of an event are either not spoken over, or a commentary is given that is articulate and will not need editing.

When editing a feature about, say, an outdoor market, the buzz of its activity is vital. While some material may be recorded in shops away from the noise of the market, probably the best material, editorially, will be acquired in the crowd. A beat pause before each question, or response, from the presenter is all that is needed to form the basis of a crossfade on atmosphere to a different section. Where the edit causes an abrupt change of character of the background then a short section of separately recorded actuality can be added to cover the change (see 'Chequerboarding' on page 102).

4.5 Recorders

The original portable acquisition format was direct cut Edison cylinders. Later, 78 rpm lacquer discs were used by reporters during WWII. They were heavy, more transportable than portable, and could not be used 'over the shoulder'. With the perfection of tape recording by the Germans during that war, the 1940s and 1950s saw the rapid appearance of more and more useful portable tape recorders.

Tape (like disc) had the advantage that the recording medium was the same as the editing medium, which was the same as the transmission medium. Removable digital media have the potential to be the same. However, at the time of writing, recorders using this sort of technology are bulky and heavy and there is no agreed standard.

Far more attractive are small recorders using Minidisc and DAT. Ordinary domestic recorders can produce superb quality at prices a fraction of professional gear. The trade-off is that these recorders use fragile domestic plugs and sockets. This is not a problem if your environment allows you to take care of the machine, or is sufficiently well financed as to regard them as disposable.

Alternatively, an option is to purchase a case designed to protect the recorder. These make the whole package larger but they have professional XLR sockets and also a large capacity battery for longer recording times. There is also the incidental, but not inconsequential, effect of making the reporter look more professional. This can make the difference between being granted an interview or rejection.

Whatever, and wherever, you are recording, you need to get the best out of your microphone and a basic review of acoustics and microphones may be helpful.

4.6 Acoustics and microphones

Acoustics

When making a recording, realize that the microphone will pick up sounds other than the ones you really want. So what does a microphone hear?

- The voice or instrument in front of it.
- Reflections from the wall.
- Noise from outside the room.
- Other instruments/voices in the room.
- General noises, if outdoors.
- Multimiking.

While it can seem that the answer to most problems is to have as many microphones as possible, in fact, the opposite is true. If you have six microphones faded up, each microphone will 'hear' one direct sound from the sound source it is set to pick up. However, it, AND THE OTHER FIVE MICS, will also get the indirect sound from its own sound source AND every other sound source.

In my example, you will be getting six lots of indirect sound for every single direct sound. As a result, multi-mic balances are very susceptible to picking up a great deal of studio acoustic.

With discussions, a lot of mics very close together on a single table can provide some extra control and give some stereo positioning. Avoid the temptation to put contributors to such discussions in a circle with comfy chairs, coffee tables and individual stand mics. The resulting acoustic quality will be excellent; unfortunately the discussion is likely to be boring and stilted. This is because once you go beyond a critical distance of about 4 feet people stop having conversations and start to make statements.

A studio designed for multi-mic music recording will have a relatively 'dead' acoustic with reflections much reduced. The disadvantage of this is that the musicians and singers cannot hear themselves properly. In a very dead studio, a singer can go hoarse in minutes! You have to substitute for the lack of acoustic by feeding an artificial one to headphones worn by the artist.

Orchestral music

Classical and orchestral music tends to be performed in a room or hall with a decent acoustic with a relatively small number of microphones; typically a main stereo mic plus some supplementary mics to fill in the sound of soloists, etc. Not only will the musicians be happier – many non-electric instruments are hard to play if wearing headphones – but also the larger number of musicians involved makes it impractical to find that number of working headphones, let alone connect them all up and ensure that they are all producing the right sound, at safe levels.

Separation

Much of the skill of getting a decent recording is to arrange good separation between the wanted sound and the unwanted sound. The classic ways of improving this are:

Booths/separate studios

Gives good separation but with the penalty of isolating the performers from each other.

Multitrack recording

Laying down one track at a time while listening to the rest on headphones gives good separation but it can be difficult to realize a sense of the excitement of performance.

Working with the microphone closer

Because the direct sound becomes louder, you can turn down the fader and therefore, apparently, reduce the amount of indirect sound from that microphone. This has the disadvantage that the microphone is much more sensitive to movement by the artist and other unwanted noises can become a problem: guitar finger noise; breathing becomes exaggerated; spittle and denture noises can become offensive.

It is a matter of opinion just how much breathing and action noises are desirable. Eliminate them totally and it no longer sounds like human beings playing the music. Too much and it becomes irritating.

Using directional microphones

Microphones are not particularly directional and are likely to have angles of pick-up of 30° to 45° away from the front.

High directivity mics will give poor results with an undisciplined performer who moves about a lot. Their main value is not the angles at which they are sensitive but the angles where they are dead. It is more critical to position a microphone so that its dead angle rejects the sound from another source. Here a few degrees of adjustment can make a lot of difference to the rejection and no audible difference to the sound you actually want.

Microphones

There is no such thing as the perfect microphone: a single mic that will be best for all purposes. While expense can be a guide, there are circumstances where a cheaper microphone might be better.

There is often a conflict between robustness and quality. As a rule, electrostatic microphones provide better quality. They have much lighter diaphragms that can respond quickly to the attack transients at the start of sounds. They have better frequency responses and lower noise figures. However, they are not particularly robust. They can be very prone to blasting and popping, especially on speech, where a moving coil microphone can be a better choice. Electrostatic microphones are reliant on some form of power supply which will either come from a battery within the mic, or from a central power supply providing 'phantom volts' down the mic lead. These fail at inconvenient times.

Quality

Yes, of course, we want quality but what do we mean by this? We want a mic that can collect the sound without distortion, popping, hissing or spluttering. With speech, there is a premium on intelligibility that is only loosely connected with how 'hi fi' the mic is. An expensive mic that is superb on orchestral strings may be a popping and blasting disaster as a speech mic.

Directivity

Directional microphones are described by their directivity pattern. This can be thought of as tracing the path of a sound source round the microphone; the source being moved nearer or further away so that the mic always hears the same level – at the dead angle the source has to be very close to the mic. At the front, it can be further away.

Beware that sound picked up around the dead angles of a microphone can sound drainpipe-like and can have a different characteristic from the front.

As a rule, the more directional a microphone is, the more sensitive it is to wind noise and 'popping' on speech. When used out of doors, heavy wind shielding is essential. Film and television crews usually use a very directional 'gun' mic with a pick-up angle of about 20° either side of centre. They are invariably used with a windshield that looks like a large furry sausage.

The most common directivity mic is the CARDIOID (Figure 4.1), so called because of its heart-shaped directivity pattern. Cardioid microphones have a useful dead angle at the back. Placing this to reduce spill is more important than ensuring that it is pointing directly at the sound it is picking up.

A so-called HYPERCARDIOID is slightly more directional (not more heart-shaped!) (Figure 4.2). It has the disadvantage that it has a small lobe of sensitivity at the back and is dead at about 10° either side of the back of the microphone.

Another common type of microphone is called OMNIDIRECTIONAL as it is sensitive equally in all directions (Figure 4.3). Omnidirectional microphones are particularly suitable for outdoor use as they are

Figure 4.1 Cardioid

Figure 4.2
Hypercardioid

Figure 4.3
Omnidirectional

Figure 4.4 Typical appearance of a barrier mic

Figure 4.5 Hemispherical

Figure 4.6 Figure-of-eight

much less sensitive to wind noise and blasting. They can be better for very close working, where directivity is not important, as they pop or blast less easily; an important consideration with speech and singing.

BARRIER mics (PZMs) can produce good results with little visibility as a result of being attached to an existing object (Figure 4.4). This can be anything from the stage at an opera, to a goal post on a football field. They work best when attached to a large surface such as a wall, floor, table or baffle. The resulting pick-up pattern is hemispherical; an omnidirectional pattern 'cut in half' (Figure 4.5).

A more specialist directivity pattern is the FIGURE-OF-EIGHT (Figure 4.6). This microphone is dead top and bottom and at the sides, but 'live' front and back. This directivity pattern is most often found in variable-pattern microphones. Figure-of-eight mics can give the best directional separation provided that they can be placed so that the back lobe does not receive any spill, or can be used to pick up sound as well. Ribbon mics have the very best of dead angles that can be very useful when miking an audience which is also being fed sound from loudspeakers.

Variable pattern mics

These, usually very expensive, microphones offer nine or so patterns. They work by actually containing two cardioids back to back (Figure 4.7).

If only one cardioid is switched on then, as you would expect, the mic behaves as a cardioid.

If both cardioids are switched on and their outputs added together, then an omnidirectional response is obtained.

If they are both switched on and their outputs subtracted then a figure-of-eight pattern results.

The other intermediate patterns are obtained by varying the relative sensitivities of the two cardioids.

Beware, these derived patterns are never as good as the same pattern produced by a fixed pattern mic. The cardioids retain their physical characteristics so an omni mic pattern derived from the two cardioids does not have the resistance to wind noise and blasting of

Two cardioids added
give omni response

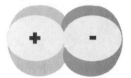

Two cardioids subtracted
give figure-of-eight
response

Figure 4.7 Variable
pattern microphone
using back-to-back
cardioids

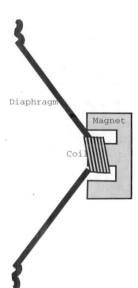

Diaphragm

Magnet

Coil

Figure 4.8

a dedicated omni mic. Most microphones are 'end-fire': you point them at the sound you want. These mics are 'side-fire' with their sensitive angles coming out of the side of the mic.

Transducers

Microphones are a type of 'transducer'; a device that converts one form of energy into another. In this case, acoustic energy is converted into electrical. Most microphones fall into two categories: electromagnetic and electrostatic.

Electromagnetic

Most electromagnetic microphones are moving coil. They are often called dynamic mics. They are like miniature loudspeakers in reverse; the acoustic energy moves a diaphragm attached to which is a coil of wire surrounded by a magnet (Figure 4.8). This causes a small audio voltage to be generated.

A specialized form of electromagnetic mic is the 'ribbon' mic that was very widely used by the BBC. These have an inherent figure-of-eight directivity. They consist of a heavy magnet and a thin corrugated 'ribbon' of aluminium which moves in the magnetic field to generate the audio voltage (these are 'side-fire' mics).

Electrostatic

Electrostatic mics, also known as 'capacitor', or 'condenser' microphones, fall into two types. Many require a 'polarizing' voltage of about 50 V for them to work. This is not a problem in a studio that can provide the necessary power as 'phantom' volts down the mic cable. Some mics contain 9 V batteries which operate a voltage boost circuit to provide their own polarizing voltage.

Electrostatic mics work by making the diaphragm part of a 'capacitor'. This consists of two plates of metal placed very close together. When a voltage is applied, a current will flow momentarily and the plate acts as a store for this charge. This is due to the attraction between positive and negative electricity across the narrow gap. If the plates are moved closer together, then this attraction increases and the plates can store more charge. If they move apart then the attraction reduces and the amount of charge that can be stored is reduced. The consequence of the plate distance changing is a current through the circuit.

If one of the 'plates' is the capsule of a microphone and the other is the diaphragm, made of a very thin film of conductive plastic, then we have the basis of a high quality microphone. The very lightness of the diaphragm gives it the ability to follow high frequencies much better than the relatively heavy assembly of a moving coil microphone. As the diaphragm moves out, so the storage of the capacitor is reduced and charge is forced outwards. As the diaphragm moves inwards, the storage increases and the current flows in the opposite direction as the capacitor is 'filled up' by the polarizing voltage.

However, particularly for battery operated mics, an 'electret' diaphragm can be used. This does not need the polarizing voltage, as a static charge is 'locked' into the diaphragm at manufacture. It is the electrostatic equivalent of a permanent magnet. While the technology is much improved, these tend to deteriorate with age.

There is a variant on the electrostatic mic where the capacitor, that is the diaphragm, is part of a radio frequency oscillator circuit. The changing capacitance changes the frequency of the oscillation, producing a frequency modulated signal that is converted into audio within the microphone. RF capacitor mics have a good reputation for standing up to hostile environments and are frequently seen as 'gun' mics used by film and TV sound recordists.

Robustness
Is the mic likely to be dropped, blown on, driven over? Does it blast on speech or singing?

Environment
Are you indoors, or outdoors? Is the mic going to get wet or blown upon? Non-directional (omnidirectional) mics are least prone to wind noise. This can

be further improved by a decent windshield. In practice, using your own body to block the wind can be the most effective.

Moving coil (dynamic) microphones are the most resistant to problems with moisture or rain. Umbrellas are of little help in rain because of the noise of the rain falling on them! The windshield is usually enough to prevent rain from getting into the mic. If you ever need to record under water, or in a shower, then a practical protection is a condom.

Size and weight
Presenters and singers do not like carrying heavy mics. Mics slung above people must be well secured by two fixings, each of which is capable of preventing the mic falling onto the audience or artists.

Visibility
This is affected by size and weight, but is also affected by the colour of the microphone. Television producers of classical music programmes want minimum visibility. This may be achieved by mics being painted 'television grey' and by small physical size.

Operas on stage are often miked by six or so PZM/barrier mics laid on the stage isolated by a layer of foam plastic (stage mice).

Power requirement
Total reliance on phantom power can mean losing everything if the power supply goes down. Batteries fail unpredictably (even if you thought you just put a new one in. In the stress of a session, it is easy to put the old one back in. Get into the habit of scratching a mark on the old one, before taking the new one out of the wrapper). Unpowered electromagnetic mics have definite advantages here.

Reliability, consistency and 'pedigree'
Microphones made by reputable manufacturers, bought through recognized dealers, are likely to be more reliable and long lasting than an unknown make bought through a dodgy outlet. They are also

likely to be consistent in quality so that one XYZ model mic sounds just like another model XYZ mic.

4.7 Multitrack

While most computers are equipped with single stereo sound cards, it is equally possible to fit cards that can handle more than two channels at a time. The main use for these is in music recording but they can also be useful for speech, drama and documentary. However, multitrack is very intensive in its use of data. To state the obvious, a CDR with 80 minutes' capacity of stereo will only have room for 20 minutes of eight-track audio.

Many multi-channel systems have quality benefits even if you rarely need multitrack recording. They use 'break-out' boxes so that the critical audio components are outside the hostile environment of the computer. These boxes can be placed conveniently on the desk while computer can be floor standing.

If a long run is not required then Minidisc, hard disk or even tape cassette-based multitrack recorders can be an option. There are also video cassette-based eight-track systems which can provide about 40 minutes' continuous running.

While there may be a temptation to record a discussion multitrack with a view to pulling out extracts for a programme, in practice it is quicker (and cheaper) to have the discussion mixed straight down to stereo by an experienced sound balancer. You also save by not having to allocate time for mixing down the material from multitrack to mono or stereo.

5

Transfer

5.1 Reviewing material

Having acquired all your material, how do you transfer it all onto your computer?

Most material will have to be copied, in real time, through the sound card of your computer, preferably using the digital inputs when the source is digital to preserve the best quality.

You should not waste this time, for it is now that you can review the material and make notes. You can save time where you know that things were redone by skipping the false takes. This is especially easy if you have a note of ident points if you were using DAT or Minidisc.

Your notes should have a convention so that they mean something to you several days, or weeks, later. Write down who the speaker is and what tape or disc they are recorded on – you did label the recordings didn't you?

Be consistent as to how you log where things are in the recording. With reel-to-reel or compact cassette, you are usually stuck with arbitrary numbers produced by the playback machine's counter. These can be wildly different on a different make of machine.

Digital machines log progress as time but there are different options. Quite a few portable machines actually record date and time of day continuously. This does mean that, on location, you can use your wristwatch and a notepad to identify points. However, a lot of mains machines do not handle this information.

DAT machines usually have different counter modes:

- counting time from when last reset
- counting absolute time from the beginning of the first recording on the cassette
- counting time starting from 00:00:00 each time a new track marker is encountered.

Minidisc recorders usually offer:

- time elapsed of track
- time left of track
- time left of disc.

Once you have established which timing system you are going to use, make your notes. Unless you have good shorthand you will not be able to transcribe what is said. Instead, log each question with the time and then list the major points made in the answer. There are many formats possible, one such is:

Joe Bloggs 29th March

0'00	*?Why has this organization been set up?*
0'10	*Fulfil need by public*
0'20	*Contact point for victims*
0'35	*Money not available from government*
0'50	*?Should public money be provided?*
0'55	*Not a practical proposition*
1'05	*Get on with it*
1'20	*500 people already involved*
1'30	*?How long before organization effective?*
1'35	*First effects within 6 months*
1'45	*90 per cent after two years*
1'55	*Complete after three years*
2'10	*OUT ... 'Everyone needs this now'.*

When you are editing something that is scripted, then most of the edits will be overlaps (going back to the beginning of the sentence at the time) or retakes (sections redone after the main recording). It should not be necessary to edit fluffs as they should be covered by overlaps and retakes. Pauses may need adjusting.

When marking an overlap mark, where it begins and how many times the speaker went back. There is nothing more irritating than editing an overlap and then discovering another one a few seconds later. Some people put brackets, the number of brackets indicating the number of repeats. You can play out at speed by holding down the mouse button on the spool button during playback. Identifying the repeats should be relatively easy, even at speed.

Better results are usually achieved by not cutting at the beginning of an overlap but cutting, instead, a few words into it, because speakers tend to overemphasize the start as they are angry with themselves for making the mistake. With this in mind when making notes, mark where, in the overlap, you think will be a good return point.

It is amazing just how easily edits can be missed. The ear is sensitive to a break in the speech rhythm. It is always a mistake for the speaker to apologize or to swear at themselves when they fluff. This has the uncanny ability to maintain the speech rhythm and it is these very edits that are most likely to be missed, especially in a news and current affairs situation. The professional way is to stop, give three beats pause and then go back to the start of the sentence without comment. This will sound more natural if the edit is missed but, ironically, that three-beat pause will usually mean that it is not missed.

5.2 Head alignment

If you are using an analogue recorder, a compact cassette machine or a reel-to-reel quarter inch tape machine, then try to use the same machine that you used to record to dub your material onto your computer. This is because analogue tape systems are very sensitive to head alignment differences between machines. This gives a muffled sound. On a cassette, Dolby C doubly emphasizes this sensitivity. In practice Dolby B noise reduction will reduce tape hiss to below the ambient noise in your recording environment without being hypercritical of head alignment.

Head alignment can be an issue with digital machines as well and can mean that a recording will play on one machine but not on another. Try always to have the original machine available, just in case.

5.3 Digital

In some ways, these recordings are the easiest to transfer. You do not have to worry about setting levels as you did that on the original recording. If you got them wrong then there is nothing that you can do to correct this until the digital audio data is on your computer. This is because the digital interface is merely transferring the numbers representing the audio that the original recorder laid down at the time of recording.

There are several standards for digitally interfacing your computer to a recorder. The most common is known as S/PDIF which stands for Sony/Philips Digital Interface. This comes in three main variants, one electrical and two optical.

Electrical

The electrical connection uses standard RCA phono plugs, as used for audio on a lot of domestic equipment (Figure 5.1). A single connection is used for both the left and the right sides of a stereo signal. Unfortunately many domestic digital recorders, while having an electrical S/PDIF input, do not have an equivalent output.

Mains operated machines will usually have an input and output in optical form but many portable machines have no digital output at all. A cynic might believe that manufacturers think that their machines will only be used for copying CDs, rather than for creative work.

For best results you should use phono leads designed for digital connections. While any old audio leads you happen to have lying around may get you out of a jam, they can lead to unpredictable problems by introducing errors. This is because they are not

Figure 5.1 RCA phono plug

designed for the supersonic frequencies involved in digital transfer. Purpose-designed digital leads tend to be fatter than audio leads.

Optical

Figure 5.2 TOSLINK optical plug

Instead of electricity, optical connectors use modulated red light to transfer the data. This light is visible to the naked eye when a connector is carrying an output. This is extremely useful in avoiding confusion between input and output leads.

There two physical standards. The original, usually found on mains operated equipment, is known as TOSLINK (from TOShiba Link; Figure 5.2). Small portable machines use optical connectors the same shape as audio 3.5 mm minijacks (Figure 5.3). Some machines have dual function sockets where the socket can either be used for audio or for optical digital. Both connectors use the same signal, which is S/PDIF in light form, so machines using different connectors can be joined using a minijack to TOSLINK lead.

Figure 5.3 'Minijack' optical plug

Some early DAT machines had a proprietary connector at the machine end and a lead terminating with a TOSLINK plug. This can lead you to find yourself wanting to copy from recorder to recorder to 'clone' a tape but are faced with joining two TOSLINK connectors. While not recommended, this can be done. Here the plastic tube containing the ink in a cheap ball-point pen can come to the rescue. Snipping the ink-free centimetre off the top will provide you with a suitable 'gender-changer'. Make sure that you trim it short enough for the ends of the TOSLINK connectors to touch (Figure 5.4).

Figure 5.4 Joining two TOSLINK plugs

Optical leads are relatively fragile in that they will fracture if bent too far, as they are made of transparent plastic not copper. They are surrounded by opaque plastic. Because the signal is conveyed by light the connection is immune to interference from electrical equipment.

SCMS and copy protection

As well as the numbers representing the audio signal, the S/PDIF signal contains extra data including various 'flags'. Normally these need not concern the user, except for the flags indicating copy protection. Literally, the system is specified so that a digital flag can be raised that will tell a well-behaved digital recorder input not to accept the data.

This all-or-nothing approach was thought to be too crude and the SCMS (Serial Copy Management System) was devised. This allows a single generation digital copy to be made. However, if an attempt is made digitally to copy that digital copy then that is prohibited. This is an attempt to reduce multiple digital cloning of commercial recordings. However, it is made to apply to your own recordings as well. This only becomes a problem when copying from an SCMS recorder to another SCMS recorder. Digital interfaces on computer sound cards do not use the system, which is ignored. Analogue copies always remain possible.

However, the original copy protect flag can be a problem when copying from a professional DAT machine to a domestic DAT. Professional machines use a variation on S/PDIF called AES/EBU which is almost, but not quite, compatible when fed to a phono plug like an S/PDIF signal. The flags are different; a professional machine will often set the flag which means copy protect to a domestic machine, but not to itself. This leads to a situation where you can copy from the domestic machine to the professional machine but not in the other direction!

CDs

CDs can be played by the Windows CD player or Cool Edit Pro has a CD player built into it. This is fine for listening to select material, but transfer is best done digitally. This is done either by using CD 'ripping' software, often supplied by the manufacturer of the CD-ROM drive, or by having your machine set up so that Windows 'sees' the audio tracks as wave files (see 'Blue bar blues' page 76).

When they were introduced, there was a lot of hype that CDs were indestructible. Most people have discovered that they are not, except compared with LPs. Unlike LPs, which will always produce something no matter how crackly, they share, with other digital media, the tendency to work perfectly or not all. While inevitably the best advice is to look after your CDs and to handle them by their edges, failed CDs can usually be recovered.

CDs play from the centre outwards. This is to allow easy compatibility between 3 inch and 5 inch discs. The beginning of the recording contains a 'table of contents'. While this is particularly heavily error corrected, it means that if the track is covered by a mark, the CD will fail even to be recognized by the machine. If this happens examine the playing surface for a mark or marks at the inside of the playing surface.

CDs slow down as they track towards the outside. This gives the digital track a constant linear velocity. The LP had a constant angular velocity, and the speed of the playing surface past the stylus decreased as it went towards the centre. This means that half way through the playing time of a full CD is not half way across the disc but further out.

If the CD is jumping and skipping a section, then this is likely to be a scratch or an opaque mark. If the CD repeatedly sticks at a section then this is likely due to a transparent mark (like jam). This bends the laser and thus confuses the player as to where it is.

Armed with this information, it soon becomes possible to work out where to look, and what to look for. Because of the nature of the digital track, scratches radiating from the centre are much less likely to cause problems compared with scratches running along the line of the track. This is why it is always strongly recommended that CDs are cleaned with radial movements rather than the instinctive rotary action.

The first emergency treatment is to wash the CD. Smear washing-up liquid over the surface and then run water over it from a tap, helping the washing-up liquid off with radial movements of your fingers. Cold or warm water is fine but it is probably best to avoid hot water: if it is too hot for your fingers then it is too hot for the CD. Shake the remains of the water off the CD and look to see if there are any obvious marks remaining. These may need individual attention, again with the washing-up liquid. Try the CD in the player (any slight remains of water will be spun off in the player).

A really persistent mark or scratch should now be identifiable. If rescue has still failed then the mark can be cleaned with brass cleaning wadding (Brasso Duraglit wadding in the UK). Remember to rub radially.

It may be that this still doesn't work for you. If your CD player's laser lens is dirty this will make it less able to correct errors. Regularly clean the lens with a special lens cleaning CD. This has little brushes on it which clean the lens as the track is changed – usually between 1 and 3. Similar cleaners are also available for Minidisc machines.

If there is still a problem then try a different make of CD player, as different manufacturers have different error correction strategies. If you are 'ripping' the audio digitally straight off the CD, see if the CD-ROM has management software that will slow down the process. A CD that will not rip at ×12 speed may be OK at ×4, ×2 or ×1. If all fails try the audio output of the CD-ROM drive. The 'View/show CD player' menu option switches on an inbuilt CD player for controlling the CD as an audio out. This can also be useful for auditioning material while you are doing other things.

5.4 Analogue

Provided care has been taken to get the levels right on the original recording, analogue transfer can be

Figure 5.5 Tip, ring and sleeve jack

Figure 5.6 XLR connectors with DIN plug for comparison

entirely satisfactory. You should make sure that your system is connected satisfactorily and well earthed. The playback should be checked for hum pick-up. This can be due to electrical equipment on the desk or to the computer itself.

Running the dubbing lead near to a power supply (including the playback machine's own) is likely to introduce hum. Running the lead near to the computer's monitor is likely to introduce not only hum but also buzzes and whistles.

Keep the dubbing lead as short as it practical, as very long runs will lead to loss of high frequencies unless you are using fully professional equipment using 'balanced' connections with tip, ring and sleeve jacks or XLR (Figures 5.5 and 5.6).

Normally you will connect the 'LINE OUT' socket of the playback machine to the 'LINE IN' socket. The most common connectors at each end are 3.5 mm minijacks (Figure 5.7). However, the better quality sound cards have 'break-out' boxes. These either sit at the end of a thick flying lead or are mounted in one of the disk drive compartments of the computer. These will have phono or quarter inch jack connectors often with a balanced connection option for use with professional machines.

3.5mm
STEREO Jack

3.5mm
MONO Jack

2.5mm
Jack for
Power supplies

Figure 5.7 Stereo and mono 3.5 mm minijacks with 2.5 mm plug for comparison

The LINE OUTs of playback machines are of a fixed level so level control has to be done on the computer. Using the headphone output instead is an option and can give excellent quality. It can also produce crackles and distortion if the volume control is dirty!

Gramophone records

Despite the popularity of CDs, there are still occasions when old LPs are the only available source of a piece of music, or even actuality. It makes sense, therefore, to get the best possible sound out of them before resorting to software solutions within the computer.

Gramophone turntables cannot just be plugged straight to a computer sound card. Not only do they have an output much lower that the line level required by the sound card, but they also need what is called RIAA equalization. This is a standardized fixed top boost and bass cut made on recording that has to be corrected at playback.

In the past this was done by the phono input of a hi-fi amp. Nowadays, many do not even possess such

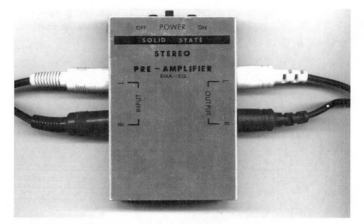

Figure 5.8 Typical record turntable preamplifier

an input. However, such a complicated device is not needed. Cheap battery-operated amplifiers are readily available; they deliver an equalized line level output suitable for your sound card. Being battery operated, they should have a lower hum level than conventional amplifiers. A typical example is illustrated in Figure 5.8. Some can double as simple mic amplifiers, if required, by switching off the RIAA equalization.

5.5 Recording

Before making a recording you have to tell the program how you want to record.

Click on 'New' and you will see a dialog asking you to set various options (Figure 5.9). This will be based on what you set previously, so, for most people, this becomes a simple matter of clicking OK.

The sampling rate is how many numbers per second are used to define your audio. For high quality use there are two main standards. 44 100 is used by CDs. If your material is going to end up on CD audio then this is the rate to choose (CD-ROM data can be any sampling frequency).

The other commonly used sampling rate is 48 000. This is required by some broadcasting companies that

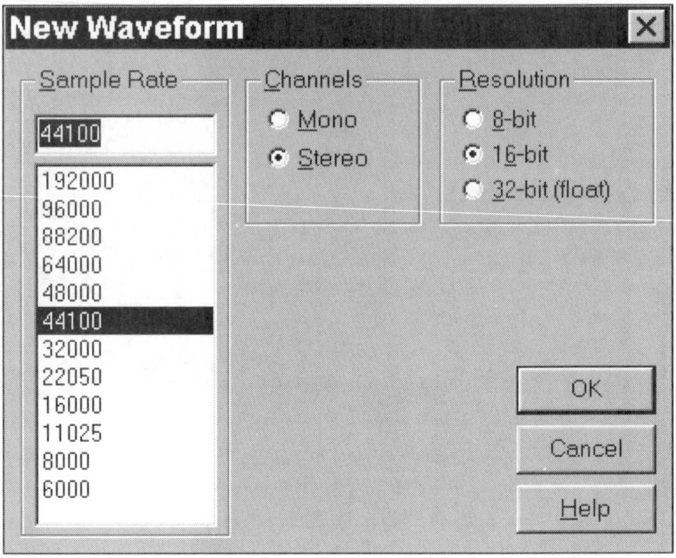

Figure 5.9

accept data files. If your material is sourced from or is going to end up on DAT then this is the frequency to use. Many non-pro DAT machines will only record analogue signals at 48 000 (or 32 000 at their Long Play speed). If you are making a digital transfer then it is vital that the sampling rate set matches the rate used on the recording. The sampling rate of a recording can be converted within Cool Edit but this is a relatively time-consuming process.

The mono/stereo option is straightforward. Operationally it is much simpler to record everything stereo, even mono sources. However, mono files are half the length of stereo files and, if everything is in mono, then make the saving. Beware that CD audio has no mono mode so burning to CD may need stereo files. However, many CD writing programs have the ability to convert mono to stereo on-the-fly, thus removing this constraint on the use of mono files. Minidisc does have a mono mode and a 74 minute Minidisc can record 148 minutes of mono, an 80 minute disc 160 minutes.

The resolution sets how big the numbers that represent your audio can be and, thus, the number of separate discrete levels that can be represented. This directly affects the noise level of your recording. Eight bit only has 256 levels and is, as a result, very hissy. It should be regarded only as a user format. With suitable control and processing 8-bit files can sound surprisingly good. However, that processing has to be done at a higher level resolution.

Sixteen-bit files are the standard, match CD and DAT and will be your normal choice. Thirty-two-bit files give the highest quality and, while taking up twice as much data space, give the highest quality results when being modified. If your acquisition source has higher than 16-bit resolution, or you are dubbing from an extremely high quality analogue source (e.g. 15/30 ips Dolby reel-to-reel tapes) then, provided that your sound card has better than 16-bit analogue to digital conversion, 32 bit will give extra quality. Similarly, if it can cope, doubling the sampling rate to 88 200 or 96 000 will further improve the quality but your resulting wave files will be enormous.

The disadvantage of working at 16 bit is that this is what the domestic consumer has at home. A 16-bit recording will have to have some headroom left in case of unexpected loud peaks. Quiet passages may have to be amplified. This can mean that your final recording is, in parts, only equivalent to a 14-bit or less recording. Working at 32 bits allows you to correct levels, normalize and then reduce to a full 16-bit master.

Setting the level

Oddly programs for audio rarely have input gain controls. The job is usually done either by a utility provided by the sound card, or by the Windows sound mixer. This is accessed by double clicking on the yellow loudspeaker icon on the task bar (single clicking produces the Windows loudspeaker volume control). This will open the playback mixer. To get at the record controls you need to click the properties item in the options menu and then click the record option (Figures 5.10, 5.11 and 5.12). You can then select which source you wish to record from.

Properties
Mixer device:
TBS Audio Mixer (6704)
Adjust volume for
Playback
Recording
Other
Voice Commands
Show the following volume controls:
Playback Controls
Wave
Synth
CD Audio
Line
OK
Cancel

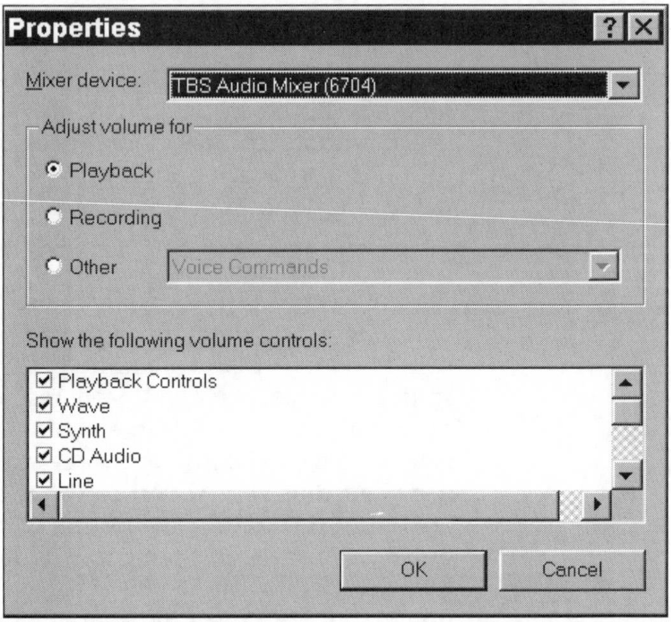

Figure 5.10 Audio mixer properties

Playback Controls
Options
Help
Properties
Advanced Controls
Exit
Synth
CD Audio
Line
Aux1
Aux2
Microphone
Monitor
Balance:
Volume:
Mute all
Mute
Advanced
TBS Audio Mixer (6704)

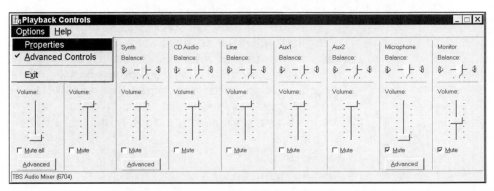

Figure 5.11 Audio mixer – playback

Recording Controls
Options
Help
Microphone
CD Audio
Line
Aux1
Aux2
Wave/Synth
Mixer Out
Balance:
Volume:
Select
Advanced
TBS Audio Mixer (6704)

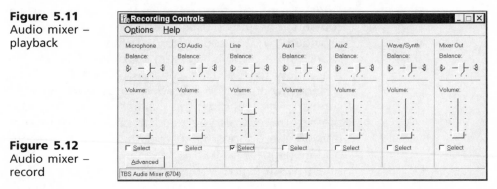

Figure 5.12 Audio mixer – record

If you have a specialist card, maybe with several separate inputs, then it is likely to have its own manager for routing the audio. Here you will have to tell the editor which card to use.

In Cool Edit this is done using the Options/settings menu (F4) and clicking on the 'Devices' tab. This will provide information on the card's capabilities and selection of the record and playback options.

Multi-channel sound cards usually present themselves to Windows as a number of separate stereo sound cards. So a card with eight analogue inputs and outputs plus a stereo digital in/out will appear as five separate cards. This is for the single wave view mode. In multitrack mode, each track can be separately selected to a sound card input or output (Figure 5.13. See Chapter 9, 'Multitrack', page 91).

Because running level meters requires a fair amount of processing time, most editors leave them off except when actually recording (playback is less of a problem as the system already 'knows' what the levels are). It is this processing overhead that makes programmers avoid emulations of analogue meter pointers like VU meters and peak programme meters.

Options/monitor record levels (F10) selects the meters in Cool Edit. The program counts this as a pseudo record mode and the red record light comes

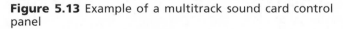

Figure 5.13 Example of a multitrack sound card control panel

on, so you can only go into record via stop. However, this visual preview gives high resolution meter monitoring. When you actually go into record, you do get a meter display but this is less precise than the preview mode; this is more than compensated for by the graphical display of your recording building on your screen.

Recording a file

Switch off the meters by clicking the stop button or using F10. Clicking the record button will switch off the meters as well but you will have to click it again actually to go into record. The meters will now reappear (but less responsive than before). The waveform of the recorded audio will be drawn on the screen in blocks. Just as the waveform is getting towards the right of the screen there will be a redraw to move the window along to keep up with the pointer which will remain static (this can be switched off if you wish).

You can crash start into record from a blank window by clicking record and pressing ENTER to the dialogue requesting the record parameters. This will use the settings that you last used.

The recording is stopped by clicking the stop button at which point the window will redraw to show you the entire waveform.

At the bottom of the screen is listed your disk resources (Figure 5.14). The first box shows the sampling rate, bit rate, mono or stereo. The next box shows how many megabytes your recording has used, followed by how many hours, minutes, seconds and frames. These two increment in blocks, rather than by the second, as Cool Edit Pro allocates the space this way. The next two boxes give you the equivalent figures but for how much space you have left.

44100 · 16-bit · Stereo	15.15 MB	0:01:27:24	1975.80 MB free	191:09.496 free	0:01:14:19

Figure 5.14 Cool Edit Pro disk usage display

How much space is available?

This raises a vital issue. What is actually meant by how much space is left? Cool Edit Pro records to its own buffer. This is different from where you intend to put the finished recording. An ideal set-up system will put Cool Edit Pro's two buffers on separate disk drives (both preferably SCSI). Finished recordings can be saved to different folders or partitions on the same disks (or, indeed, to other drives on the system). Personally I have two SCSI drives which are used as audio work files with IDE drives used for storage and non-audio programs. I also set Windows to put its TEMP file on one of the SCSI drives as this gives speed benefits for every other program on the machine. To change the Windows TEMP file destination two lines have to be added to the AUTOEXEC.BAT file:

```
SET TEMP=G:\TEMP
SET TMP=G:\TEMP
```

where 'G' should be replaced by the letter of the drive you want to use).

Topping and tailing

Once the recording has been finished, top and tail it by removing any extra material at the front and end. Ends usually sound cleaner if you use a fade out transform on the last half second or so. This guarantees that your recording ends on absolute silence and removes a potential source of clicks.

6

Editing

6.1 Introduction

Cool Edit Pro is two editors in one, linear and non-linear. They are simply toggled between by clicking the icon at the top left-hand corner of the screen. It changes appearance depending which mode it will enter.

When it shows a stylized view of three tracks, you are in the linear single wave view mode and clicking it will take you to the non-linear multitrack mode. It will then change to show a representation of a single wave indicating that clicking it will return you to the linear mode. See Figures 6.1 and 6.2.

We'll look first at the linear editor which physically changes the audio files so that data is changed when the files are saved.

When you first run Cool Edit Pro, you see a blank centre screen surrounded by a large number of icons, as shown in Figure 6.3. This should not be a cause for panic! The icons at the top can be ignored until you are familiar enough with the editor to find them

Figure 6.1 Cool Edit Pro non-linear editor icon

Figure 6.2 Cool Edit Pro linear editor icon

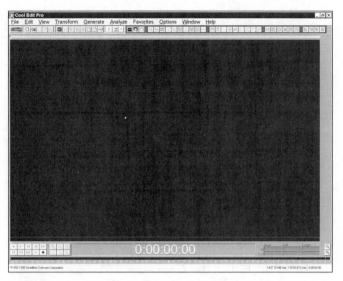

Figure 6.3

useful. The most important controls are the transport and zoom icons at the bottom. We will come to them as soon as we have loaded a file.

6.2 Loading a file

OK, maybe you have not recorded anything yet, but you will find that your computer already has a number of wave files. They have the file extension WAV (if you have set Windows to show extensions).

You will find a lot of them in your C:\WINDOWS\MEDIA directory. These are the sounds that Windows can make when it is asked to undertake various actions. You may have turned some, or all, of these off, but the wave files are still there. Indeed you may decide that your first exercise with the audio editor is to make some new sounds to personalize your computer. Loading JUNGLE WINDOWS EXIT.WAV (a clap of thunder) gives the illustrated display (Figure 6.4). There are probably as many different installations of Windows as there are machines running it. While the file I have chosen is

Figure 6.4

available on both Windows 95 and Windows 98 CDs, it is possible that it was not copied to your hard disk when it was installed. This depends on what sound scheme options were selected. It is part of the Jungle Sound scheme. In Win 98 the filename has been changed to JUNGLEWE.WAV. The length is very slightly different owing to the non-audio data being slightly different.

In Figure 6.4 the status line at the bottom of the screen reveals that the recording uses 324 k of disk space, is 3 seconds and 19 frames long, stereo, at 22 050 samples per second (CEP's display in the illustration is set to the European TV standard of 25 frames per second. Don't confuse this with the compact disc digital structure which has 75 frames per second).

This particular recording looks as if it is low level but, in fact, it is only 0.11 dB short of full modulation. How do I know this? Because I can get Cool Edit to tell me!

Selecting Transform/Amplitude/Amplify (Figure 6.5) will give the amplify dialog (Figure 6.6). You will see the DC Bias and Normalization options in the bottom

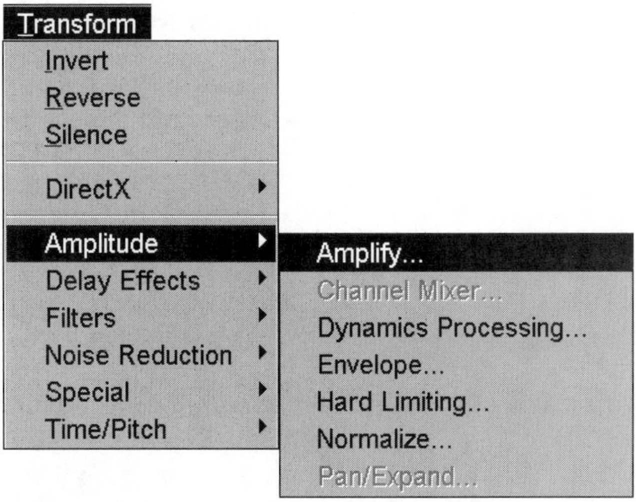

Transform

Invert
Reverse
Silence

DirectX ▶

Amplitude ▶   Amplify...
Delay Effects ▶   Channel Mixer...
Filters ▶   Dynamics Processing...
Noise Reduction ▶   Envelope...
Special ▶   Hard Limiting...
Time/Pitch ▶   Normalize...
    Pan/Expand...

Figure 6.5

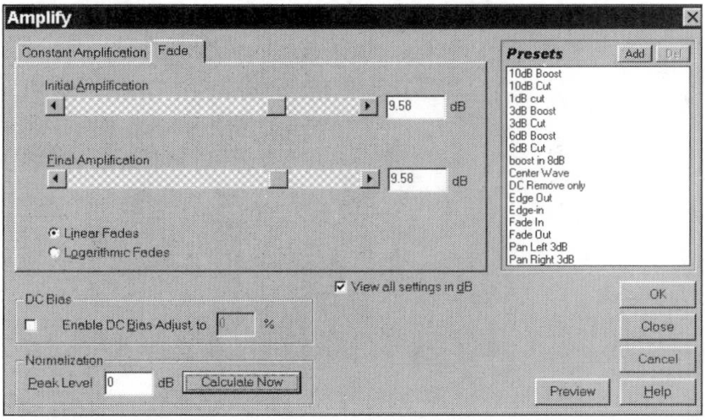

Figure 6.6

left-hand corner. Clicking the calculate button shows the amount of boost required for full normalization.

In practice, most of the time, you do not need all the options of the amplify dialog and can use the Normalize option (Figure 6.7). This allows you to set the normalization level, choose whether to apply DC Bias correction and, for stereo files, choose whether you want left and right channels separately normalized or to keep the level change the same for both

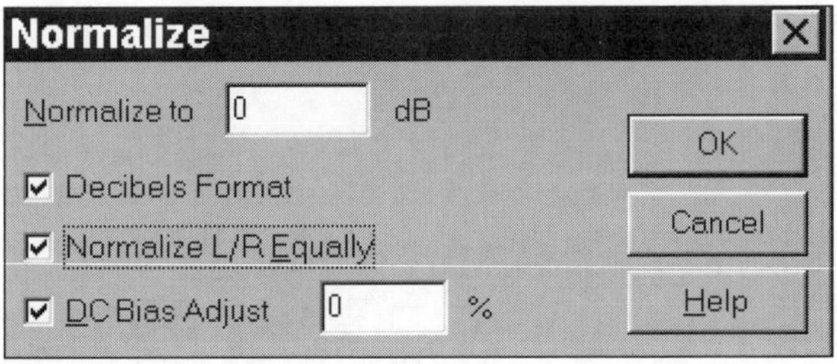

Normalize ⊠

Normalize to |0 | dB

☑ Decibels Format

☑ Normalize L/R Equally

☑ DC Bias Adjust |0 | %

OK

Cancel

Help

Figure 6.7

(this preserves sound images positions in the sound stage).

Close examination of some waveforms can reveal that they have what is called a DC offset. I have simulated a 10 per cent offset in Figure 6.8. If you look at the detail, you can see that the whole waveform is offset slightly below the centre line. Offsets like this are highly undesirable as they will produce a click every time the track is started or stopped and can cause clicks on edits.

A good quality sound card should not introduce an offset, but a lot of the cheaper ones do. This is not a disaster, as the offset can be removed by the editor.

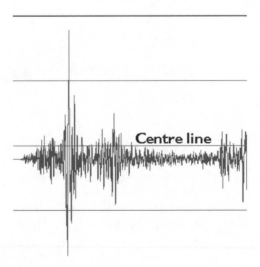

Figure 6.8 DC offset

In Cool Edit Pro this can be done as part of the normalization process when the level is corrected to produce full level. It has also an option to remove DC on-the-fly during recording.

If you actually want a DC offset you can change the value from 0 per cent to the value you require. This may seem an unlikely need but could be useful to correct an offset manually. Cool Edit Pro's DC correction introduces a bass cut. If your recording has a deliberately extended bass response you would not want to lose this (however, a sound card poor enough to have a DC offset is unlikely to have an extended response).

The Decibels format option sets the values entered into the 'Normalize to' box to show decibels rather than percentages.

The reason the audio looks so low level is that there is a brief moment at the start of the thunder where the level virtually reaches the peak. By zooming in on the detail we can see that it is literally one cycle of audio from negative to positive. Because the stereo sound is not coming from the centre, the right-hand channel (bottom display) is even lower in level (Figure 6.9).

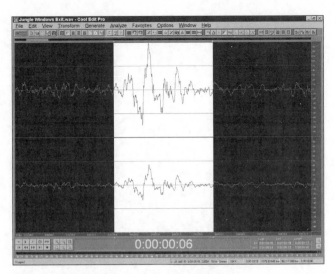

Figure 6.9 Thunder peak selected

Short peaks are a common problem with level adjustment. It is one of the prime skills of mastering engineers in the record industry to modify these so that the resulting CD can be made louder.

In this case, we can make a louder file by highlighting the section, reducing it by 6 dB and then renormalizing. We can also, if we wish, untick the Normalize L/R Equally box, to bring up the right-hand channel. Using the Edit/Zero crossings options can ensure that our selection does not stop, or start, in the middle of a cycle, which can introduce a click (Figures 6.10 to 6.12)

By doing this manipulation we achieve a louder recording, although the peak level remains the same. However, we have brought up the background noise as well, which may or may not be a good idea. You may like to try this, if you have this file on your hard disk. The process has now revealed two more peaks that are holding down the level. When do you stop? That's where skilled ears come in.

While a full level recording is ideal for production use, this may not always be the case for the version heard by the end user. Windows sound effects are a case in point; it is generally desirable to set these 6 dB

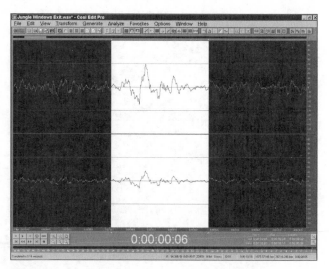

Figure 6.10 Thunder; peak reduced by 6 dB

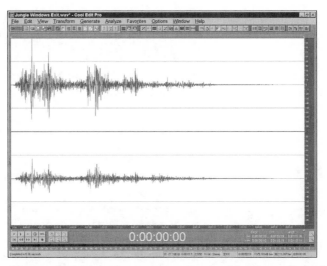

Figure 6.11 Thunder; renormalized with 'Normalize L/R Equally' set

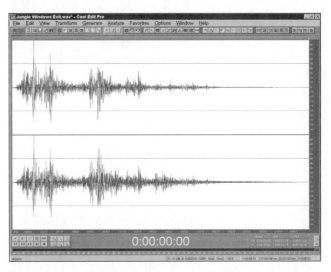

Figure 6.12 Thunder; renormalized with 'Normalize L/R Equally' not set

to 12 dB below maximum recording level. This allows the PC speakers to be set at a good level for when they are reproducing music from a CD – or even your editing session, without the Windows sound effects blasting you out of the chair!

Figure 6.13 Transport icons

To modify recordings, we need to be able to navigate round a sound file. Cool Edit Pro always presents the whole of the file when loaded. Personally I find this reassuring and have never got on with other editors that only show the beginning of a loaded file.

The keys to navigation are the transport and zoom icons at the bottom left of the screen (Figure 6.13). The left-hand block is for the transport and is reassuringly like a cassette recorder, with a few extra options. Stop, Play and Pause are followed by the triangular play icon in a circle. The last icon has an infinity sign. Clicking the standard play triangle symbol plays the file, as you would expect. However, it will stop when the playback cursor reaches the end of the screen, or the selection, if one has been made. In our present case this will be the end of the file. But if we were zoomed in, it would not be.

The triangle in a circle play button will also start the playback but this will continue to the end of the file. The last icon sets up a loop mode so that playback never ends but loops round, either to the start of the file, or to the start of the selection.

The second row of icons behave like those on a CD player. The double arrow buttons 'spool' the cursor along the file (audibly if in the play mode). The other spool buttons take you to the beginning or end of the file, or to the next marker if you are using them. The final button, with the 'red LED', is the record button.

The playback 'head' is represented by a vertical line. This can be moved to where you want, merely by single clicking the point with the mouse. If you are already playing the file then a cursor line will appear where you clicked but playback will not be interrupted. Clicking the play triangle will instantly move the playback to the newly selected point. The space bar toggles playback mode on and off. If you are using this, then it toggles through stop, before going to the new location on the next press.

You can select an area by dragging the mouse over the area you want. The alternative is to click one end, with the left mouse button, and then click the other end with the right mouse button. The selected area can be modified with right mouse clicks. The selected edge nearest the mouse click is the one modified.

The right-hand icons control zooming. The top three are absolute controls. '+' zooms in. '–' zooms out. This only affects the horizontal magnification. If you have a wheel mouse, then the wheel will also control zooming. The final icon restores the view to the whole of the file.

The bottom three icons of this block deal with the selected area of the file. The first fills the window with the selected area. The second '[' icon zooms in on the left edge of the selected area. The third ']', correspondingly, zooms in on the right-hand edge of the selected area. This is particularly useful for fine trimming the start and finish of edit points.

There are two more zoom icons at the bottom right of the window. These '+' and '–' buttons control the vertical magnification of the audio. This is not needed very often but can be useful to examine the character of low level sounds such as noise.

Holding the control key down allows zooming with the arrow keys. The left and right arrow keys control horizontal zooming (based on the centre of the screen). The up and down arrow keys control vertical zooming.

It is very easy, quickly to examine a section of audio. Click on the point you are interested in. Define a selected area by dragging or using the right mouse button and click on the 'Zoom to selection' icon. You can repeat this as many times as you like to get into really fine detail. You can even set the magnification to be so large that you can see the individual samples (Figure 6.14). These are represented as dots. They are joined by a line generated by the computer. This line is not simply dot-to-dot for cosmetic reasons, but will show what will happen to the audio after the output filter.

If necessary you can move the dots, literally to redraw the waveform. This can be the only way to

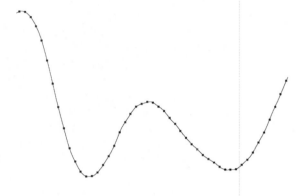

Figure 6.14 Waveform zoomed-in to show samples

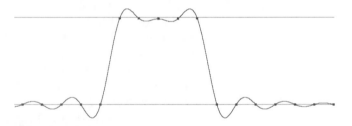

Figure 6.15 Five sample square pulse drawn onto waveform, showing computer's interpretation of the result

eliminate a very recalcitrant click that has slipped through declicking software.

In Figure 6.15 I have drawn a five sample click this way, to illustrate that the line joining the dots is more intelligent than it first appears; the wavy lines show what is known by audio engineers as 'ringing'.

6.3 Making an edit

Editing on a PC is best done visually. Some editors have a 'scrub' editing facility where the sound can be rocked backwards and forwards as on a reel-to-reel tape machine. People used to quarter inch tape ask for this, probably out of fear, or sentiment. In practice, nearly everybody edits visually. The technique is very simple and the eye and the ear work together easily. It is much easier to learn, from scratch, than is scrub editing.

Editing audio is surprisingly similar to editing text in a word processor. If you use a word processor, then you already know the basics of editing audio. You use your mouse to select a start and finish point. The space between is highlighted by reversing the colours. Pressing DELETE removes the highlighted section.

The selected section can be created in two ways. The most obvious is to place the mouse at the start of the section you wish to cut and drag the pointer to the end of the cut section. Cool Edit Pro also allows you to select the other end of the selection from the cursor by right clicking on the point (this works forwards or backwards).

6.4 Visual editing

This is all very well but how do you find a phrase within an item? Here you use the zoom facility already described.

Figure 6.16 shows a short section of speech 'The um start of the sentence'. The text has been added

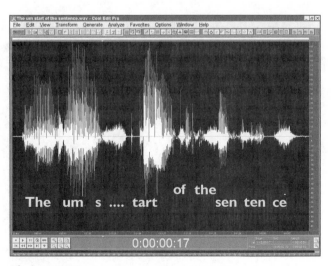

Figure 6.16

to the picture and is not a feature of current audio editors!

How do we go about removing the 'um'? Clicking the play button with the mouse (or pressing the space bar) will play the section. A vertical line cursor will traverse the screen showing where the playback is coming from. Your eye can identify the section that corresponds to 'um'.

With the mouse click at the start of this section. The cursor will move to this point.

Press play again and the playback will start from where you clicked. You can hear if you have selected the right place as, if you have not, there will be something before the 'um', or it will be clipped. Reclicking to correct is very fast.

Now identify, not where the 'um' ends, but where the next word starts – the 's' of 'start'. This time click with the RIGHT mouse button. This sets the end of the section to cut (Figure 6.17).

Hit play again and just the highlighted section will be played. Is the end of the 'um' clipped? Can you hear the start of the 's'? If not then the edit is probably OK.

Now press the delete key. The highlighted section is removed (Figure 6.18).

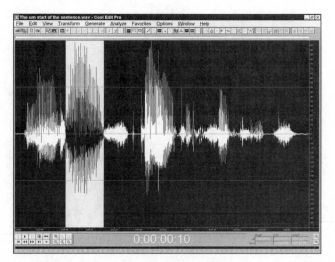

Figure 6.17 'Um' selected for removal

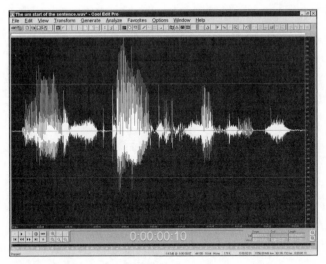

Figure 6.18 'Um' removed

Click a little before the join and play the edit.

Is it OK? If so go on to the next edit.

If it is not, then click UNDO and you will be restored to the previously highlighted set-up.

You can adjust the edges of the selected area, by right clicking near them.

At first, you may well have to cycle round several times, but soon you will be getting edits right first time. Remember even redoing the edit three or four times is still likely to be quicker than it used to take with a razor blade.

7

Quarrying material

7.1 Introduction

Returning to base with material of exactly the right duration and needing no editing is a rare occurrence. If your recording is of a continuous event then simply cutting it to time is all that is required.

With unscripted speech-based items, you can return with substantially more than you will ever use, especially if you have recorded a series of interviews which you plan to 'quarry' out as extracts, rather than use them complete.

However, it cannot be stressed too much that the very best way to edit an interview is to have asked the right questions in the first place! A well-structured interview will also be easier to edit for duration. However, circumstances can force returning with far too much material, especially if vox pop (in the street interviews) have been recorded.

Once upon a time there was no alternative; quarter inch tape had been used and the only way to edit was with a razor blade, wax pencil and sticky tape. Copy editing added extra time and was avoided.

7.2 Blue bar blues

No matter how fast your computer and its hard drives, audio files consume a vast amount of data compared with a word processor, and take time to be saved. Most editors have a very useful UNDO facility. This works by saving a copy of the section of the file you are changing onto hard disk.

All this takes time and slows down the editing process. This means that you should structure your material so as to avoid long audio files of more than 3–5 minutes. Put plainly a single edit within a 30 minute file takes much longer to do because of the time taken to save that long file. A 5 minute file takes one-sixth as long to save. A strategy of one or two files per item within a longer piece will speed things along nicely.

Extracting sections of a file to new separate files is most easily achieved by highlighting the section and using the 'File/Save Selection' menu option.

Where you just want to copy and paste, then use the clipboard where, as with all modern operating systems and programs, Control/C will save the selected area. Control/X will cut it out and Control/V will paste it to the current cursor position.

However, there will be things you will want to do to long files which will take time, such as noise reducing an archive recording. It is inevitable that you will spend some time staring at the screen as the blue progress bar slides slowly from 0 to 100 per cent. Ironically the time is much less than used to be taken with the mechanics of razor blade editing. However, it feels much longer because you, personally, are not doing anything while the process is taking place.

You can improve your efficiency by having a number of other tasks that you can do while the processing is going on; telephone calls to make, facts to check, or even letters to write, as the processing can carry on in the background while you use your word processor on the same PC.

Some routine operations, such as normalization, can often be done using 'batch' files. These allow you to leave the computer to get on with the task while you do something else.

You will very often want to load many files at the same time. Please remember that the Windows file selector allows you to select many files which will all be loaded when you click OK.

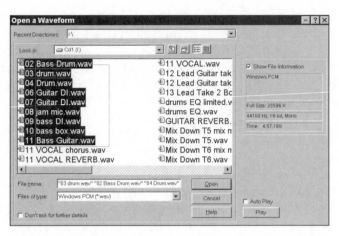

Figure 7.1 File selection using 'rubber band'

One way is to use the 'rubber band' method of multiple selection (Figure 7.1). The mouse point is placed by a track and the left mouse button pressed and held down. With the left button still held down, moving the pointer will produce a rectangular 'rubber band' which will select any files included within it. You can include all the available files by starting the rubber band at the first file and then dragging downwards and rightwards. The file selector window will scroll the list so that you can include everything.

If you want to select a number of individual files then click on each file you want (Figure 7.2). Normally selecting a new file will deselect the previous selection, but if you hold the control key down while you do this then this will not happen. The control key also introduces a toggle action so that clicking on a selected file will deselect it. This is particularly useful if you want to make a minor modification to a number of files as, once loaded, switching between them is rapid. Once you have made all the changes you can SAVE ALL and all your changes will be preserved.

Obviously, loading and saving a large number of files takes an appreciable time but by triggering this with one action you can be doing something else that

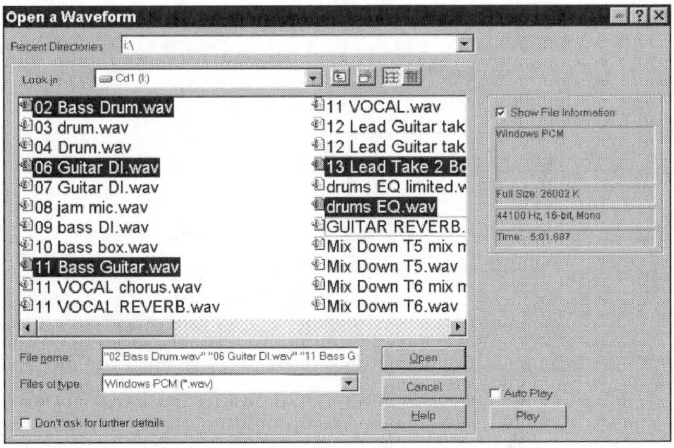

Figure 7.2 Individual file selection using mouse with control key held down

is useful while this is happening. If you are saving 10 files each taking 15 seconds to transfer you have given yourself 2½ minutes to make a phone call or whatever.

Many programmes will include material from compact disc. This can be commercial music but also sound effects, or even from your own personal archive. Here considerable time can be saved as a properly set-up computer will be able to 'rip' the audio data from the CD much faster than real time. This means that 10 minutes of material can be transferred is less than 1 minute. You don't even have to listen to it! The quality will be better as the audio is not converted back to analogue and then back to digital again by your sound card.

The easiest way is to use a background program that modifies the Windows desktop so that audio CDs appear in the desktop window with their tracks showing as .WAV files. These are usually named 'track1.wav', 'track2.wav', etc. This allows you to copy the audio files as if they are normal data files. Later versions of Windows come with this feature built in. Earlier versions can have this added either by using a special file, or a utility provided by the manufacturer of the CD-ROM drive. Windows desktops without this feature will open a CD drive containing audio, but the tracks will be shown as 'TRACK1.CDA' etc. They are not directly accessible as audio.

7.3 Copy, cut and paste

Cool Edit Pro gives you the option of six separate clipboards. It has five of its own, plus the standard Windows clipboard. These are selected, whether through the 'Edit/Set current clipboard' menu option or by using control/1 to control/5 for Cool Edit Pro's own clipboards and control/6 for the Windows one.

This means you can work with multiple pieces of audio 'in memory' at the same time; so you can, for example, copy different jingles or link music sections to each clipboard, and place them in your file at chosen locations. The current clipboard can also be set to be the Windows clipboard. This is available to other programs and is a convenient way to copy audio from Cool Edit Pro to another program or vice versa.

These internal clipboards save audio in your temporary directory as wave files, and they can be retained even after Cool Edit Pro closes. The 'Delete clipboard files on exit' setting in Options/Settings/System sets switches this on or off.

As well as using the normal paste function to insert material you can create a new file from the clipboard using Edit/Paste To New (Shift/Control/N). Save selection will usually be quicker as you do not have to copy to the clipboard first.

Edit/Mix Paste (Shift/Control/V) will add material to the file on top of existing audio; the length is not changed except if the insert option is used. If the format of the waveform data on the clipboard differs from the format of the file it is being pasted into, Cool Edit Pro converts it before pasting.

There are four mix paste modes:

- *Insert*, inserts the clipboard at the current location or selection, replacing any selected data. If no selection has been made, Cool Edit Pro inserts clipboard material at the cursor location, moving any existing data to the end of the inserted material.

- *Overlap*, the clipboard wave does not replace the currently highlighted selection, but is mixed at the selected volume with the current waveform. If the clipboard waveform is longer than the current selection, the waveform will continue beyond the selection.
- *Replace*, will paste the contents of the clipboard starting at the cursor location, and replace the existing material thereafter for the duration of the clipboard data. For example, pasting 1 second of material will replace the 1 second after the cursor with the contents of the clipboard.
- *Modulate*, modulates the clipboard data with the current waveform. I suspect that this option is here because they can do it. It is yet another way of producing weird noises.

8

Structuring material

8.1 Standardize format

Use the linear editor to prepare material for mixing in the non-linear multitrack editor. You need to fix internal levels, correcting them as required, and check for various technical parameters.

It makes sense to standardize on your format. Sixteen bit will be quite adequate for most speech-based items. Thirty-two bit adds complication but can give substantial quality improvement where a lot of processing has to be done.

The major decision is whether to use 44.1 kHz or 48 kHz (or their multiples 88.2 kHz and 96 kHz). If your work is going to be distributed on CD then 44.1 kHz makes total sense. Some broadcasters require 48 kHz. This can also be convenient if you are originating and distributing on DAT as only professional machines will record analogue inputs at 44.1 kHz (they will record at this sampling rate if fed digitally from external equipment). Remember that if you are working at 48 kHz you will have to convert any CD recordings that you use. For domestic and semi-professional use, 44.1 kHz is usually the most convenient.

If you have not done so already, now is the time to split your audio material into sections. The basic rule is that every time there is a change that will need to be adjusted split the audio into separate files. Local differences, like a mumbled sentence, can be handled within the linear editor but, when it is likely that there will be a need for a crossfade, then split the file.

Each section should be 'Edged-in' and 'Edged-out'. Internally the levels should be adjusted so that each file enters the multitrack non-linear stage with its levels adjusted and with equalization or processing applied as required.

Optional cuts should be left in. They can be made in the non-linear editor. This will mean that they will stay in their correct place in the wave file and can easily be restored if the item turns out to be, that rarest of happenings, short.

8.2 Useful 'Favorites'

You will often need to match the levels of sections and it is worth the trouble to put some of the common basic operations into the 'Favorites' drop-down. Some suggested useful options are:

Edge-in
Within the amplitude/amplify transform option set up the fade tab for an initial amplification of –3 dB and a final amplification of 0 dB.

Edge-out
Within the amplitude/amplify transform option set up the fade tab for an initial amplification of 0 dB and a final amplification of –3 dB.

These two 'edge' options will fix most level changes. There is nothing to prevent you using the option more than once for severe differences. This will usually be quicker than going into the amplify transform option. Edge-in and -out are also useful to smooth the first and last second of an insert. The listener's ear will not be aware of the level change except as being smoother than it might otherwise have been.

To use, select the beginning of the level change (usually an edit) and extend the selection a second or so to where the new level is established. The 'edge' functions then change the level at the edit and then fade back to the level of the insert. This should be consistent overall as it will be the same recording.

Level changes owing to a change of recording are best handled using separate wave files within the multitrack mixer.

3 dB cut/3 dB boost/6 dB boost/6 dB cut

These options will cover most level correction problems where a whole section needs to be matched. A typical cause is a retake where the recording level has not been reset properly.

Fade out/Fade in

These are vital where an insert has a heavy background atmosphere. It is usually convenient to put these fades into the inserts using the linear editor but there is also the option of performing fades as edit decisions within the non-linear editor section. If the background is typical of a non-busy room then a fade out in the linear editor is best. The function of the fade in or fade out is to prevent the listener being aware that there is any background noise.

However, if the background is traffic noise, children in a swimming bath, an airport, etc., then it is important to have several seconds of noise before the start and at the end of the wanted material. At the beginning the noise will be 'tucked under' the link or previous item. At the end it will be lost under the next item, or used as an actual fade, to provide a 'passage of time'.

(As an aside, it is vital to identify a heavy background behind an item especially if it is not relevant to the item: 'With the road being dug up outside his office, the mayor explained the council's education policy.' Otherwise the listener will be distracted by the background, rather than hearing what is said.)

Normalize 0 dB/Normalize –6 dB

These set the level of a section so that the highest level is set to the peak level available (0 dB) or at –6 dB. The second option is mainly useful when mixing down multitrack music as this can give you more headroom if you are using 16-bit mixdown files rather than 32 bit.

Overall normalization is best done on all the audio files in one go using a batch file, as you can be doing something more useful than watching the progress of a blue bar while this is being done. However, you sometimes find that a section of an interview is quieter than the rest – the interviewee is answering an embarrassing question perhaps. Here it can be useful to re-normalize just this section. If necessary, the top and tail of the section can be matched using the edge-in and edge-out options.

It is also useful to have some additional preset options within the amplify transform as shown in the diagram. These can be used 'straight' or as the starting point for a one-off setting (centre wave is a do-nothing option for when you want to start from scratch).

Figure 8.1
Amplify presets

8.3 Batch files

Cool Edit Pro allows you to set up batch files to automate some activities. Using batch files is quite an advanced activity and they will take a while to get right. This means that, in the short term, they can add to the time taken to edit an item. However, once they are sorted, they can be great aids to efficiency. Batch files can be grouped into three types.

The first type 'starts from scratch' and creates a new file. The first command has to be File/new (Cool Edit Pro will impose this if you forget). This would be useful if you had a standard set-up you wanted to create regularly. An example might be to create a file at a particular sampling rate and bit resolution. Generate 20 seconds of line-up tone followed by 10 seconds of silence.

The second type is where the same sequence of treatment is applied to the whole file and you can select a series of files to be actioned. This means that you can run a series of transforms that takes several hours to do in total but, instead of several hours of blue bar blues, you can be off doing something more interesting, such as lunch.

The last type of batch file is one that is going to be applied to a highlighted section of a file and consists of a series of treatments; say, equalization followed by normalization followed by 6 dB of hard limiting.

Scripts and batch processing can be found under the Options menu heading.

The scripts are, literally, text files containing instructions in a sort of programming language. But don't panic! You create scripts, not by writing them, but by getting Cool Edit Pro to record what you do to a file. You can save this and recall it in the future.

To use an existing batch file or script, click on its name and then click 'Run Script' or 'Batch Run' (Figure 8.2). The 'Batch Run' button will only be enabled if the script is designed to operate on the whole file.

To record your own script is simply a matter of entering a new name in the New Script title box. This will enable the record button. Which type of script you generate is controlled by your starting point in the editor. If there is no file then the 'Start from Scratch' mode is assumed. If a section of the file is

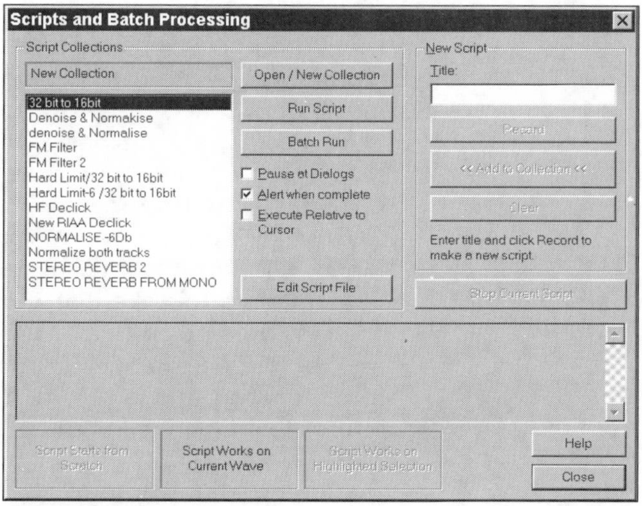

Figure 8.2 Scripts and batch processing dialog showing typical range of scripts

highlighted then a script for action on a highlighted section is created. A file with no selection creates the sort of script that can be applied to a series of different files.

Clicking 'record' starts the process. You now perform the actions you want to be able to repeat each time the script is run. If you had ticked the 'Pause at Dialogs' button then when the script is run, the dialogs will appear for you to enter settings (or press Enter to OK them). With the box not ticked then the values you enter as you record the script are used. When setting up a script, it makes sense to use a short test file, so you do not have long to wait for each action to be processed during the record process.

The 'Execute Relative to Cursor' is an advanced option and is used when operating 'Works on Current Wave' script types. With this, you can have all script operations performed offset from your cursor position. If you record the script with the cursor set 10 seconds in from the beginning, then, when the script is run it will operate 10 seconds after your current cursor position. Recording with the cursor at zero is a way of producing a script that works from the current cursor position, rather than using a highlighted section. This might be used for generating line-up tone sequences, for example.

As you would expect, the script recording is stopped by clicking 'Stop current script'. The '<Add to Collection<' button is now enabled so that you can save what you have done, or you can cast it to oblivion by clicking 'Clear'. This is the best option if you have made a mistake. For most people who are not Techies, redoing the recording is preferable to clicking the 'Edit Script File' button, as this dumps you into Notepad and a raw text file that is the actual script of commands for the whole collection of batch files.

Running the scripts in the future is a matter of going to the same dialog, selecting the sequence you want and clicking 'Run Script' for action on the current file, or 'Run Batch' if you want to process a number of files.

The Batch Process dialog will appear. Click 'Add files' to get the file selector and add the files you want to the list (don't forget that the Windows file selector allows you to select more than one file at a time). If, having loaded a number of files, you see some that you realize you should not have included then you can highlight them with the mouse and remove them by clicking the 'Remove' button.

The directory box sets the path where the modified files will be put. This allows you to keep the originals, if you have enough hard disk space. You can use different file formats and they will all be saved in the format set in the 'Output format' box.

The Output Filename Template is an odd, but powerful, option if you can get your brain round it.

The names of files in your batch can be modified before being saved to the Destination Directory. When running a batch, the processed file's extension will automatically change to that of the format chosen in Output Format (e.g. *.AIF for Apple AIFF). You can, however, force another extension, or alter the filename itself (the portion before the '.'), by using the filename template. There are two characters to use when altering the Output Filename Template: a question mark '?' will signify that a character does not change. A star, '*', will denote the entire original filename or entire original file extension.

Here are some examples of how filenames (taken from the help file) will be saved given the original filename and the filename template:

Filename	Template	Result
Zippy.aif	*.wav	Zippy.wav
Toads.pcm	Q*.voc	qtoads.voc
Funny.out	B???????.*	bunny.out
Biglong.wav	????.wav	bigl.wav
Bart.wav	*x.wav	bartx.wav

The scan list button will check all the files and tell what types you have selected. This can be useful in picking up a rogue file that has, somehow, appeared in the wrong format. The Change Options button allows you to define how 32-bit data are saved.

The two tick boxes at the bottom give important options. 'Disable UNDO' does just that. This gives a great speed advantage and is not even dangerous if you are saving the processed files to a different directory. The 'Overwrite existing files' box must be ticked if this is what you want to be able to do, otherwise the file will be skipped.

9

Multitrack

9.1 Loading

The multitrack editor can look very intimidating, as in Figure 9.1, with a full-scale multitrack music balance shown. However, on initial loading a much simpler appearance is shown, as in Figure 9.2.

Beside each of the four tracks shown are the individual controls for each track, as shown in Figure 9.3.

The essential ability of a multitrack recorder is that it can record on all of its tracks simultaneously, on only one, or on any number in between. So, in

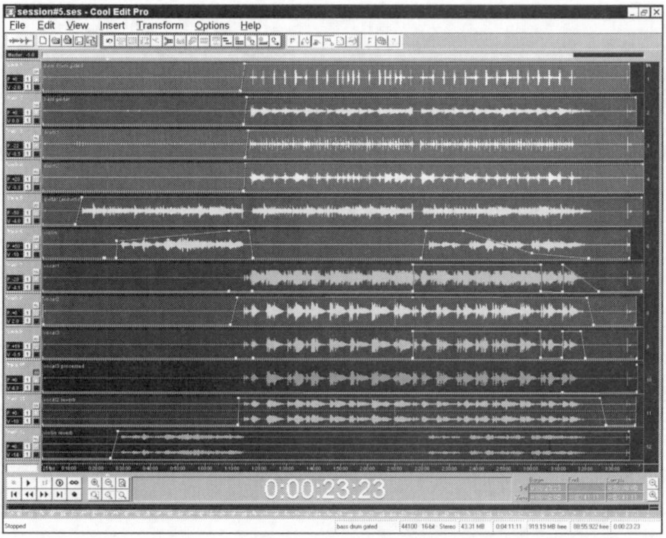

Figure 9.1 Multitrack music mix down

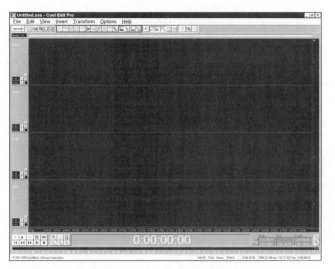

Figure 9.2 Multitrack opening screen

Figure 9.3 Track
controls

addition to the transport record button, each track has a red record enable button marked 'r'. If this is not selected then the track is in 'safe' mode and will not go into record when the main record button is selected.

With multiple tracks, you are not restricted to stereo recording. Cool Edit Pro can record up to 64 tracks simultaneously, provided that the computer is equipped with enough sound card inputs and has fast enough processing and hard drive capability.

The two numbered squares allow you to select each track to individual sound card inputs.

For playback, each track, be it mono or stereo, is fed to a stereo output on the selected sound card (Figure 9.4). A mono track is selected to only one output, by panning fully left (odd numbered output), or fully right (even numbered output). Although this may seem odd, this is fully in line with the convention used on multitrack music mixing consoles.

The record input can be fed from the left (odd) or right (even) card input or in stereo (Figure 9.5).

When starting a 'New session', you can also select between 16- and 32-bit recording. The sampling rate is fixed for the entire multitrack session. The 16/32-bit

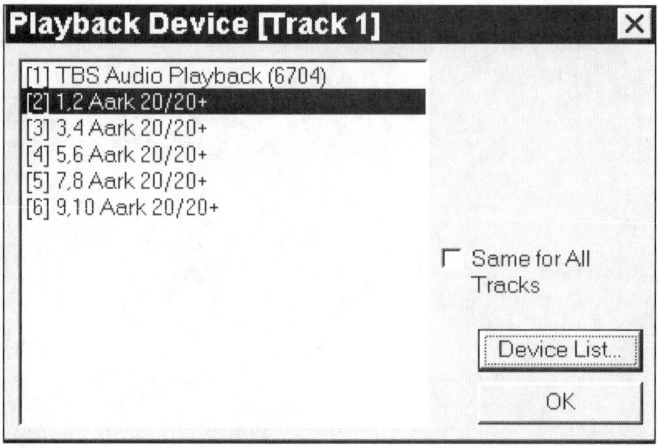

Playback Device [Track 1] ☒

[1] TBS Audio Playback (6704)
[2] 1,2 Aark 20/20+
[3] 3,4 Aark 20/20+
[4] 5,6 Aark 20/20+
[5] 7,8 Aark 20/20+
[6] 9,10 Aark 20/20+

☐ Same for All
Tracks

Device List...

OK

Figure 9.4 Selecting playback outputs

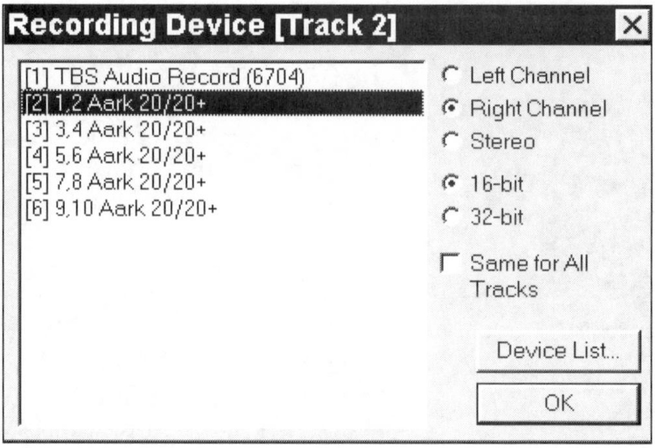

Recording Device [Track 2] ☒

[1] TBS Audio Record (6704)
[2] 1,2 Aark 20/20+
[3] 3,4 Aark 20/20+
[4] 5,6 Aark 20/20+
[5] 7,8 Aark 20/20+
[6] 9,10 Aark 20/20+

○ Left Channel
◉ Right Channel
○ Stereo

◉ 16-bit
○ 32-bit

☐ Same for All
Tracks

Device List...

OK

Figure 9.5 Selecting record inputs

option in the New Session dialog (Figure 9.6) selects what the mix down output will be. You can mix 16- and 32-bit wave files simultaneously. The greyed out mono/stereo options are not used.

Important though multiple ins and outs are, they are likely to be little used outside music mixing. The multiplicity of outputs are useful mainly for feeds to an external mixer. For us, the power of the non-linear editor is the ease with which it can merge and mix our mono and stereo audio into a continuous mono

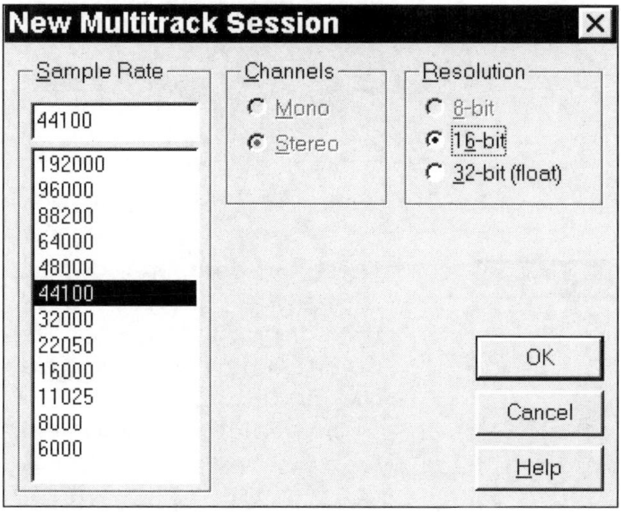

Figure 9.6 'New session' dialog

or stereo final mix. Our audio items are added and laid out, as required, on any number up to 64 of mono or stereo tracks. They can be regarded as a large number of play-in machines in a studio. Without our intervention, all the material will be mixed together at full level with no panning.

It is time to look again at those tiny squares on the left of each track. We have already described the record button, the record and the playback selects. The two squares labelled with 'm' and 's' are mainly useful in music balancing.

The 'm' mutes a track. This can be useful for switching off a track without losing all of its settings.

The 's' button is the solo facility. When selected, this mutes all the tracks except that one, and any other tracks also with their solos selected. Their pan settings are maintained and so this is equivalent to 'solo in place' on a multitrack mixing console.

9.2 Pan and volume

The two vital boxes are the 'P' and 'V' windows for panning (Figure 9.7) and track volume (Figure 9.8).

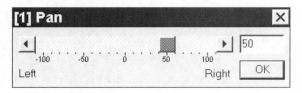

Figure 9.7 Track pan control

Figure 9.8 Track volume control

These can be operated in three ways.

One is to place the mouse cursor over the box and then drag it up, or down, to change the setting. The other two involve using a pop-up obtained by double clicking on the box.

The pop-ups each have a slider that can be dragged by the mouse, or moved by holding down the left-hand mouse button with the cursor placed on the slider on the side you want the 'knob' to move.

The third alternative is to edit the value in the box. Positive pan values can be up to 100 per cent (fully right). Negative values can go down to –100 per cent (fully left). You can pop up as many of these sliders as you like and place them where you like, virtually building yourself a mixer if you wanted. These set the overall level, and pan, of the track so that every item on it is affected by this.

As well as these 'global' settings for the whole track, you can control Volume and Pan on a moment-by-moment basis. This is done using 'envelope' controls within the track. To be able to see the envelopes, you have to set the 'View/Show pan envelopes' and 'View/Show volume envelopes' menu options so that they are ticked.

Until changed, the Pan envelope is a straight line down the centre of the track, and the Volume envelope is a straight line along the top of the track. These represent Pan Centre and Full Volume (as modified by the track settings on the left). Clicking on a line will produce a small blob where you clicked. This is a 'handle' which can be dragged by the mouse to change the setting.

To produce a simple fade-out, create a handle where you want the fade to start, and another one to the right of it. Drag this second handle to the bottom of the track (representing zero volume) and left or right to the point you wish the fade to finish. (You will not be allowed to drag it to before the start of the fade handle. However, this can be a useful 'end stop', if you just want to switch off the track.)

You will discover that the level immediately begins to increase after the fade, because the end of the envelope line is attached to the handle at the end of the track. This needs to be dragged to the bottom, so that the whole of the rest of the track is faded out.

Now you have your basic fade you can adjust it for timing by listening and, if necessary, moving the handles. A simple straight line fade is often inadequate but you can add as many intermediate handles as you like to make the fade curved, or to have several sections (see section on 'Fades and edges', on page 103).

You only need to use the pan envelopes where you want the sound from a track actually to be heard moving. You should beware of doing this to speech, as it is a distraction and reduces communication; the listener is likely to react on the lines of 'Oh they've started moving. Why? ... er ... what did they say?'

The pan envelopes are set in entirely the same way as the volume envelopes although, much of the time, a single pan setting for the overall track is all that is needed.

In both cases, a handle can be removed by dragging it off the top or the bottom of the track.

Figure 9.9 shows a short section of a mono track. This starts at full volume and ends with a fade. The pan setting starts at centre. Then there is a sudden pan

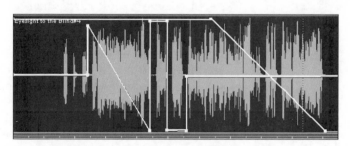

Figure 9.9 Example of using envelope controls for pan and volume

full left followed by a slow pan from full left to full right. The next two sections are panned fully right and then fully left with the track ending panned centre.

When you want to move a handle, you position the mouse cursor over the handle until the cursor becomes a pointing finger. You can now drag the handle to where you want it, positioning it anywhere between the previous and next handles. It takes a little while to get used to this, and it is very easy to create new handles rather than moving an existing one. UNDO can correct this, or dragging the handle off the track will remove it.

This use of envelopes is a common way of overcoming the imposition of 'one-finger' operation by using a mouse. While you can only change one thing at a time, all the changes are remembered exactly. While simpler and less impressive than a screen filled with a pretty picture of a sound mixer, it can be much more effective.

To get rid of one-finger operation you need either an external mixer or an external control device that may look like a mixer, but controls the software settings with the program. Many Digital Audio Workstations work this way, but at a price.

9.3 FX tracks

You now have complete control of the audio levels, and timing, of your material. Some editors allow real time processing of audio, applying equalization and special effects such as reverberation. This requires a lot of processing. Cool Edit Pro does not provide this, but some sound cards have Digital Signal Processors which take the load off the computer's processor and have software to allow effects to be added in real time. Where this is not possible, the technique is to make additional tracks which contain the effects.

For example, to add reverberation to a soloist's voice: load the soloist track – which is likely to be mono – convert it to a stereo track, and save it with a different name such as 'Soloist Reverb'. Now run a reverb transform on a short section of your new track. Adjust the settings until you have the sound you want. Now set the 'dry' sound to nothing and adjust the 'wet' sound so that it modulates the track at a sensible level. Now transform the whole track. This reverb track can now be added in with the original when mixing (see section on 'Reverberation and echo' on page 134).

9.4 Multitrack for a simple mix

It is very easy to see the non-linear multitrack mode as a complication to be avoided. For simple cut and paste editing, this is often true; however, as soon as mixing becomes involved, it is invaluable. It gives you the ease of multi-machine mixing with quarter inch tape, combined with absolute repeatability and flexibility.

Because of the integrated nature of the editor it can be convenient to 'pop in' to the multitrack editors when all that is required is to make a simple mix.

The following example takes a section of Berlioz's *Damnation of Faust* which has one bar repeated in it. A simple cut edit does not work as a retake has been cut in, which does not match the original. The answer is to cross-mix between the two takes during the one bar overlap.

This is simply done by entering the multitrack mixer. If the wave file is already in the linear editor, it will already be in the editor's waveform list (F9). All that needs to be done is to highlight it, and click the Insert button (Figure 9.10). (To

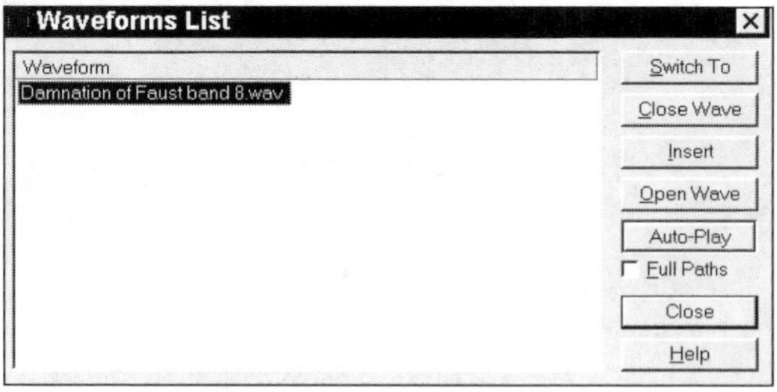

Figure 9.10 'Faust' file to be inserted

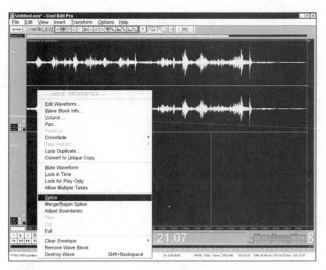

Figure 9.11 'Splice': really split, not join

make the illustrations clearer, I have zoomed in vertically to show just two tracks.)

Next, the end of the first take of repeated section is selected with the mouse, and the track split at that point. This done by right clicking on one side of the cursor and selecting the option to split the track (Figure 9.11). Annoyingly, Cool Edit Pro calls this 'splice' which, in most parts of the world, means join rather than the opposite.

Now that the audio is in two parts, the second half can be dragged, using the right mouse button, down to track two. In Figure 9.12 I have highlighted the first take of the repeated section.

The next thing to be done is to slide the second section to the left until the overlapping section matches musically that of the track above (Figure 9.13).

At this stage, the rhythm matches but the level bumps and a crossfade needs to be implemented. This can be done manually with the volume envelope but it is much easier to use the automatic crossfade facility.

First the crossfade area is highlighted. Then both sections are highlighted by clicking with the shift key

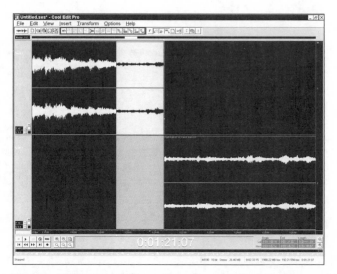

Figure 9.12 'Faust' file split and second section moved to track 2

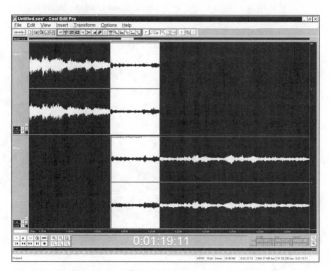

Figure 9.13 'Faust' slid to correct timing, ready for cross-fade

held down. Now select the crossfade from the menu (Figures 9.14 and 9.15).

There are four types of crossfade to choose from (Figures 9.16–9.19). They all have their uses. However,

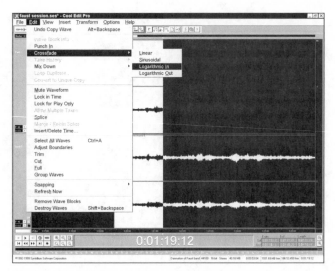

Figure 9.14 Selecting the crossfade option

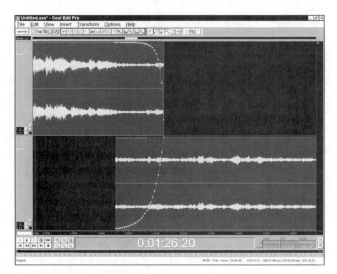

Figure 9.15 Crossfade has been implemented

linear crossfades tend to give a boost to the level in the middle of the fade. Logarithmic fades should give a better approximation to equal power thoughout the transition. Because of their asymmetrical shape, there is a choice of 'direction'. Sinusoidal is another equal power

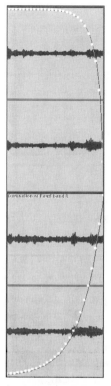

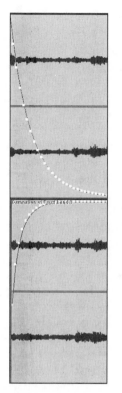

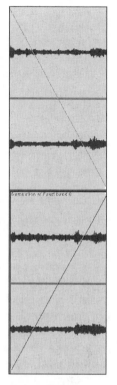

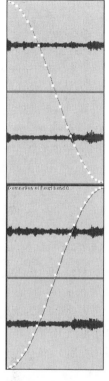

Figure 9.16
Crossfade (log in)

Figure 9.17 Cross-
fade (log out)

Figure 9.18
Crossfade (linear)

Figure 9.19
Crossfade (sine)

option. There is no substitute for trying the different options. The UNDO option is one menu click away (or ALT/BACKSPACE). Playing the different transitions will soon convince you as to which is best for any edit.

To transfer this edit back to a file in the linear editor you merely mix down your tracks and then save the file. If you are confident with your editing, you will select a short section with a little each side of the crossfade and mix down just that section. This can be copied and pasted in to your original in seconds.

9.5 Chequerboarding

Once a programme item becomes much more than a simple interview or talk, then mixing becomes necessary. While mixing is possible with linear editors using Mix Paste, it usually involves a lot of UNDO cycles to get it right. The multitrack non-linear editor makes it so easy.

In the days of reel-to-reel tape, programmes of any complexity would be split into banded 'A' and 'B' reels with odd numbered inserts on the 'A' reel and even numbered inserts on the 'B' reel. Each insert could then be started over the fading atmosphere at the end of the preceding insert. Where both items had heavy atmosphere the incoming band would start with atmosphere and a crossfade made. Where necessary a third 'C' reel would be made up to carry continuity of atmosphere over inserts with tightly cut INs and OUTs.

In a multitrack editor, this technique is often known as chequerboarding. Figure 9.20 shows a programme of music linked by a presenter. Some of the music is segued and so the chequerboard principle is invaluable. For convenience, the links are given their own track (track 1) and the music tracks their own 'A and B reels'. (Segue is an Italian music term adopted by radio and show business to mean one item following another without a break. It is pronounced 'seg-way'.)

A degree of extravagance in allocating tracks is not a problem. In the days of reel-to-reel tape, we were

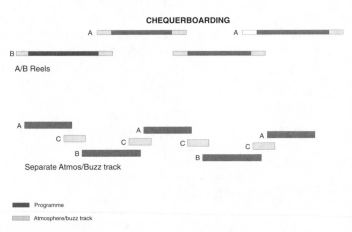

Figure 9.20 Reel-to-reel chequerboarding

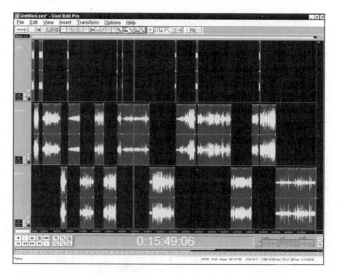

Figure 9.21 Chequerboarding on Cool Edit Pro

often limited by the number of physical play-in tape machines available (and by how good the tape-operator was!). Here, each track is effectively a different play-in machine. We can have at least 64 of them – rather better than a four-machine radio studio (Figure 9.21).

Even better, we are no longer reliant on the tape operator. Each 'play-in' is automated; fixed until we change it. No longer does one muffed play-in require a whole take to be redone. If an item comes in slightly late or early, it only takes a moment with the mouse to slide the track until it is right.

While totally dominant in music recording, reel-to-reel multitrack was never a very useful device for drama and features because of this one limitation; you could not slide tracks relative to one another. There are ways to emulate this, but only by adding considerable complexity to the process.

9.6 Fades and edges

However, a little more than simple chequerboarding is required. Together, a studio manager on the mixing console and a good tape operator would make material merge seamlessly by skilled and instinctive use of faders, both on the console and on the play-in machines. Often inserts would be started, not with an actual fade-in, but with edge-ins, where the item was started with a sharp fade-up to half fader, and then a fade-in over a second. Sound balancers have also learned that most items sound better if they are started 2 dB to 4 dB low and edged up, over a second or so.

On a PC, all these instinctive skills have now to be defined. This is why many people, with the necessary money, will pay for an external mixing device to control the computer's mixer. This preserves the power minutely to adjust levels within an item merely by having your fingers on the faders. This 'pinky-power' can control up to ten sources simultaneously. A computer mouse can only control one. However, because of the automation provided by a PC, each change can be made individually and cumulatively.

It makes sense to preserve the old split of duties between the sound mixer and the tape operator. When preparing your material, pay attention to how each file begins and ends; put a rapid fade-out on the atmosphere at the very end of the file. Put an edge-in on the front. You can set this up as a 'Favorite'. Starting values of −6 dB and −3 dB fading up to zero are useful here.

Edge-ins should be inaudible to the listener. They are there to smooth the seams. Actual fade-ins and fade-outs − ones that are going to be perceived as such by the listener − are best done using the envelope controls of the multitrack mixer.

Fade-ins usually have the role of establishing a passage of time. The old convention, from the 'golden age of the wireless', was a slow fade-out followed by a slow fade-in. These days, the fades are much faster and the fade-in is more of a steep edge-in.

Fades also have the function of allowing voice-overs to be heard over music. Very rarely should the listener be aware that this is being done. What will sound particularly horrible will be a mix where the music is audibly dipped to make a hole for the voice to enter, with another hole following the end of speech before the music is faded back up.

Most important is that the voice-over should fit the music. The music should return at the start of a musical sentence. The dip-down is less critical in the context of pop music but, with more formal features, the voice should pick up from a cadence, or the end of the musical sentence.

When mixing voice-overs 'live', I used to say to presenters that I would 'go on the whites of your tonsils'. By this I meant that the music would dip on the very first syllable of their link. Equally it sounds better if the fade back up starts about a syllable before the end of the link. This is less critical if the presenter has hit the beginning of a musical phrase as you are fading up in the micro-pause between notes.

Music fade-outs are often done badly. A vital thing to realize is that any fade-out consists of three separate parts (Figure 9.22):

1 warning the listener they are about to lose the music
2 fading down
3 fade-out.

This applies as much to a clean fade as to one where the presenter comes in over the end.

The classic example of the latter is the opening signature tune, a feature of so many audio items,

Notional structure of a fade-out

Warning   Fade   Fade-out

Figure 9.22

from broadcast programmes to cassette promotions. The music starts full level and then towards the end of a musical phrase, the level is dipped slightly to warn the listener and then almost immediately after faded down so that the presenter can start their link at the end of the phrase with the music running under. The music is then lost under the presenter, being out on a cadence, or end of phrase.

Fading classical music badly can cause real offence, but pop music deserves equal care. Again the three-part fade applies; a slight dip at the start of the last phrase and then a fade towards the closing cadence, with the music taken out at the end of the cadence. A common mistake – especially when the fade is made while looking at the score – is to leave the start of the fade too late so that the listener is not prepared for the music to end.

Where the fade has to be quick, because the illustration is about the words of a song, then, rather than lose the last few words on a fade to cadence, a 4–6 dB dip at the beginning of the last line can provide the psychological warning and make the quick exit far less offensive.

9.7 Prefaded and backtimed music

A common technique to bringing a programme out on time – especially a live one – is to have 'prefaded' closing music. A piece of music is dubbed off, say of 1 minute duration, and started with the fader shut exactly 1 minute before the end of the programme. As the presenter makes their closing remarks, the music is faded up behind them, to bring the programme to a rousing and punctual close.

A technique used by some producers is to end the last item with music which is then run under the presenter's close and then faded up so that it ends. This effectively offsets the prefade back to the last item. However, it sounds so much better if that last piece of music actually begins at a musically satisfying place, rather than just emerging arbitrarily. A cheat that works most of the time – although I remain offended that it does – is to dub off the item's closing music to the end, and run that separately as a backtimed prefade closing music. The two recordings will not be in synch, yet, because they are the same music, they can be crossfaded under the presenter's closing link without being audible to the listener.

These techniques may seem unnecessary for a recorded programme, as it can be edited to time. However, time to edit can be a scarce resource and it may be much more cost-effective to use a prefaded closing music than to spend another hour looking for cuts.

9.8 Transitions

Transitions are the stuff of dramatic effect, not only in drama itself, but also in documentary. Crossfading from one acoustic environment to another or from one treatment to another can be extraordinarily effective.

The first type of transition, going from one location to another, is simply managed. Two tracks are used and the incoming track slid along to overlap the end of the exiting track. You can use Cool Edit Pro's automatic crossfade generation, as described earlier in this chapter, or you can manually create the envelopes if this does not produce the effect you want. Always remember that you can modify the crossfade that Cool Edit Pro did for you by moving the envelope 'handles' with the mouse. With a transition, you are rarely going to use a perfectly engineered equal loudness crossfade.

The second type of transition is from one type of treatment to another on the same material. Examples include an orchestra playing in another room. The music is treated to remove all the high frequencies. The characters move into that room, and you want the high frequencies to return as they make that move, or they open the door.

Sometimes you want to go from music or speech apparently heard acoustically – a juke box playing in a coffee bar – to music heard directly off the recording. Here you need to take the good quality original and treat it, literally by playing it on a loudspeaker and recording it with a microphone (or, in a studio, playing the music on the foldback loudspeaker).

With the example of the orchestra, your treated track is identical in length with the original and is inherently in synch with it. Within the multitrack editor place the cursor roughly where you want the treated track and, without moving the cursor, insert it. They will both line up on the cursor. Now they need to be grouped together so that you can move them without losing synchronization. Click the first track so that it is selected and any other selection lost. Now, holding down the control button, click the other. They should now both be selected. Right click on one of the waves and click group waves on the pop-up. You can do this from the 'Edit/Group waves' menu item as well. Now, when you move them about they will move together, until you match the dialogue in the best way.

You now take the handles at each end of the volume envelope on the good quality track and move them to the bottom, so that there is no output from the track. Listening to the mixed output, adjust the track gain for the treated track so that it fits behind the dialogue. If it is supposed to be coming from behind a door, then you may wish to pan the track to one side or the other.

At the point in the dialogue where you wish to make the transition, create a crossfade by adjusting the envelopes on both waveforms. You can adjust the gain of the incoming untreated track with the track volume control, or by using the volume envelope.

In the second example, the music recorded off a loudspeaker, the two files will be of different lengths. However, it is very easy to slide one file with respect to the other until you can hear that they are in synch. Now you can group the two files and treat the transition in the same way as before.

9.9 Multitrack music

Music recorded on a multitrack recorder can be transferred to your computer very easily, if you have a multitrack sound card with enough inputs. This can even be done digitally with the right gear. A common interface is an variant on the optical TOSLINK connector. The same plug is used, but an eight-channel digital signal is sent instead. An increasingly common type of multitrack recorder is the video cassette-based machine, which records eight digital audio tracks instead of a picture. These can be linked together to become 16-, 24-, 32-track etc. recorders.

In cases of desperation, short pieces of music – say up to 5 minutes – can be transferred using only stereo equipment. The multitrack tape must be prepared by having a synchronizing 'slate' recorded simultaneously on all of the tracks just before the music. A convenient way of doing this, in a studio, is to use the slate talkback: the talkback that goes to the multitrack and not to the artists. A short ident, and a sharp sound like a coin tapping the talkback mic, should be adequate. At the end of the music, a similar slate is recorded. The tracks can now be copied two at a time, either directly to the computer, or to a stereo recorder, preferably digital. When the tracks have been transferred to your computer, they should be split into their corresponding mono files. This will have been done automatically if you have used the multitrack facility to record them as two mono files rather than a single stereo file. Each wave file will begin with identical recordings of the slate. Each file is edited so that it starts with the most recognizable part of the 'sharp' sound. Sample accuracy editing is advisable, and practical, here.

Once loaded into the multitrack editor and lined up at 0:00:00:00 the files should now be back in synch. How much error has been introduced can be ascertained by examining the end slates. Usually, in my experience, the error, even with an analogue multitrack, is better than 2 EBU frames (2/25th second) which for many types of music (and performance) is good enough. Better accuracy can be obtain by using Cool Edit Pro's transforms to varispeed tracks digitally so that the error is corrected. The mix down can now be undertaken within Cool Edit Pro.

10

Production

10.1 Introduction

This is where the disparate strands of material you have been gathering get pulled together and made into your final programme, or programme item. The script is written or a running order prepared. Each item is put into the correct order, checked with the existing material and recorded. Traditionally this was done as a studio session. While there are still plenty of occasions when this is still the case, the digital audio editor can make this less necessary.

With many speech-based programmes, the main purpose of the studio is to provide a quiet, and relatively dead, environment for your presenter to record their voice links between items (aka inserts) and voice-overs on top of actuality. Given that your presenter is in this sterile environment, it makes sense for them to be able to hear the insert material in context so that they can react to it.

However, small, cheap and light, portable recorders make the studio increasingly unnecessary. Its remaining function of providing a sound console seamlessly to mix the material together is subverted by the digital audio editor.

10.2 Types of programme

There are many types of recorded programme, from a simple talk to a full-scale drama or music recording. A modern PC has within it the resources to cope with all of these. The only programmes not within its scope are, arguably, those that need two-way communications. The interview on the telephone or circuit is still best done in a studio with its specialist communications and talkbacks. While a PC can handle a modern ISDN line, it is the sophisticated communication between the producer and the interviewer, as well as the two-way discussion on the circuit, that is important. While it is not impossible for a computer to do this, the requirement is sufficiently specialized as not to be something that you can buy easily.

Incidentally, when time is available, there is a technique for recording telephones in good quality even when international calls are involved. Put simply, each end of the conversation is recorded locally, in good quality, on a portable recorder or, of course, in a studio. The two tapes, or discs, are then combined in the digital editor after they have been posted, or freighted, to bring the two halves together. With digital recorders, little slippage of synch will take place but, even when it does, a little sliding of tracks can easily be done. A track can always be split while it is silent because 'its' person is listening to the other end. The main difficulty with this technique is that people tend to speak differently on the phone. On a poor line they shout. This sounds very odd when both voices sound as if they are in the same room. Incidentally, the best results are achieved if the room atmospheres at each end are kept running continuously.

10.3 Talks

The most straightforward, if not easiest, production is the straight talk. Once common on radio, this now struggles to keep in view. However, it also manifests itself as commercial promotions – a sales message on cassette or, even, the chairman's morale boosting address to the company's employees.

This merges into the illustrated talk which can soon become a magazine programme.

From the production point of view talks tend to be serial in nature. They start at the beginning and finish at the end. Talks will be all the more effective if they are not overburdened with detail. The BBC Radio Talks Department, in the 1950s, thought that the ideal talk would handle one concept in 15 minutes, provided that the structure was of the form:

In this talk I shall tell you X.
I am telling you X.
I have just told you X.

This may be effective in communicating X but it will be difficult for it to be entertaining. Somewhere between will be what you want. Do you want the audience to retain the facts, or to get a good impression of your chairman?

There are two likely locations for this talk; a special recording or one made at a public meeting. In the first case, you are in control. In the second, you are not. Do not be tempted to use your best music mics, as they are almost certain to be a popping disaster. A good moving coil mic that is known to be insensitive to making popping and blasting noises is best. It is also unlikely to have an extended bass response that will pick up every low frequency distraction, from distant slamming doors to air conditioning noises.

Will the speaker stand or sit? When recording a public meeting, ignore what the speaker's office tells you: use two microphones. One should be set up for the speaker sitting down. The other should be rigged for the speaker standing up. If at all possible, use floor stands for the mics so as not to pick up table thumps. Place the mics about 2 feet (60 cm) from the speaker's head. Any closer and the balance will become very sensitive to head and body movements. The further away they are, the less likely the speaker is to move them (there is a perception problem here; if a mic is very close and it is pointing at their lips, they have an urge to move it so that it is 'pointing at them'. This leaves the mic pointing at their eyes. In fact, this is not a problem, only their desire to move the mic is. 'Miking' the eyes can, therefore, make sense).

If a mixer is not available, then splitters can be purchased which will allow the two mono mics to be fed separately, one to the left-hand channel and the other to the right-hand channel of a portable recorder.

If you want audience reactions – applause, etc. – then you need an additional pair of mics left and right of the audience. These should be fed to the mixer, if you are using one, or to a second recorder (two Minidisc recorders are more generally useful and cheaper than a decent mixer). If both of the recorders are digital then there will be no problem with synchronizing the recordings once transferred to the computer.

For the specially recorded talk, you are more in control. Someone sitting at their desk, with a table stand holding the microphone, will produce adequate results from an audio point of view but, very likely, produce a very formal delivery. This may be what is required.

Another option is to record the talk as if you were recording an interview, but without any questions being asked. The speaker has eye contact and is more likely to talk to you, representing the idealized single listener. The vital thing to remember with all recorded (and broadcast material) is that you are communicating with individuals. There may be thousands of them but they are not a mass audience; they are a large collection of individuals listening by themselves.

10.4 Illustrated talks

Best results are obtained if the illustrations can be played in as the talk is recorded. This allows the speaker to react to the material more naturally than if they record 'cold'. Whether or not the inserts, as played in, are recorded at the time is down to production convenience. A minimalist set-up might be two DAT or Minidisc recorders; one being the play-in machine, perhaps feeding a couple of small powered speakers as used for PC sound.

10.5 Magazine programmes

Structurally, the magazine programme is a series of premixed items linked by, usually, one or two presenters. In many ways it is rather like an illustrated talk but on a larger scale. The individual items are often 'presented' by reporters and contributors other than the main programme presenters.

10.6 Magazine item

Magazine items often take the form of illustrated talks with a presenter/insert/presenter/insert format but they can also be simple features.

10.7 Simple feature

A simple feature differs from an illustrated talk by being continuous with an attempt to make the item seamless. The presenter will talk over actuality, and interviews will be chopped up and presented as extracts.

10.8 Multi-layer feature

The multi-layer feature is the most complex form. Various threads are interwoven to make a seamless whole. Before computer audio editing these were time consuming, labour intensive and complex to make. The multitrack non-linear editor doesn't remove the complexity but makes it much more easily handled. Expect at times to be simultaneously mixing a dozen tracks.

10.9 Drama

Drama brings with it the widest range of techniques. It can be formal and theatrical when reproducing a stage play. It can also be shot like a film using portable equipment on location.

Drama is often studio bound for the simple reason that, with no cameras, the actors are traditionally not expected to learn their lines but to read them off scripts. There is often a technician in the studio to do 'spot' effects such as pouring tea and knocking on doors.

Each speech in a drama script should have a left-hand margin with the character name at the start of each speech which should also be numbered, starting from 1 on each page. This allows speedy and simple communication by referring to 'Cue *N* on Page *Y*'.

Drama often, but not inevitably, leads to a lot of post-production. Background effects are best done at the time. It is especially valuable to the actors if they can hear what they are talking against. This is the advantage of location drama, as the sounds are real. If the actors are on a street, they raise their voices quite naturally as a bus passes. This is also a problem with location drama, as you will probably have no control over that bus passing. It can also result in tiring the listener, unless there are contrasting scenes away from the hurly-burly.

Modern technology allows drama, set in modern times, to be made quite effectively with portable equipment and a PC for post-production. The biggest problem can be making it clear to passers-by what is happening. People are now quite blasé about camera crews but can be quite confused when there is no camera. It has even been suggested that a balsa wood mock-up camera would be an asset, to prevent your best take being ruined by the microphone picking up passers-by speculating, amongst themselves, as to what all these strange people are doing!

10.10 Music

This encompasses the whole range from the solo singer, to the multitracked band, to the symphony orchestra. There is also the range from electronically synthesized and sampled music, usually using MIDI, to acoustically produced sound from traditional instruments. Music technology is a whole industry and is beyond the scope of this book except to say that, surprisingly, little has to be added to a PC already equipped for what we are describing, to take on the full gamut of music.

11

Post-production

11.1 Timing

This covers the period between the compilation of an item – often in a studio – and the final mastering of the finished item. In some cases, it will be difficult separately to distinguish this phase, but it is rare for it not to exist at all.

With a 'live' transmission, it is rare for adjustments not to have to be made after a rehearsal. Only in live news and current affairs programmes do adjustments have to be made on-air. An overall studio producer/director will ask colleagues to cut arbitrary amounts of time out of items. Hellish though this can be, it usually works.

Typically, when all the sections of an item have been put together, whether in a studio or with your audio editor, the item will be too long, poorly paced, with further editing of sections required either for style and pace or 'fluffs' and overlaps. It is always a waste of time to 'de-um' speech recordings until this stage as otherwise you waste time on edits that are thrown away.

You should not regard it as a 'mistake' when you find your post-production material is too long. This is because it is difficult to judge the worth of individual items until they are in their final context. Something that seemed relevant within an individual item becomes a drag within a programme. Ten per cent overrecording is about right. Now is also the time to be rigorous with yourself.

A good policy when looking for something to cut is, to quote Jack Dillon of the 1950s BBC Radio

Features Department, 'Cut your darlings'. That piece which was particularly difficult to get, that fact that is so fascinating to you, will, as likely as not, turn out to be supremely indulgent and weaken an otherwise tightly crafted programme.

The post-production phase is also where you apply your audio design, which is covered in the next chapter.

11.2 Level matching

It is important that levels are matched as the human ear is extremely sensitive to quick changes in audio levels. It is much less sensitive to slow changes of level. This is, at once, a danger and a boon.

The danger is that, unless you are constantly keeping an eye on levels, they can slowly sink. There is a tendency for the ear to want successive sounds to be about 2 dB quieter for the level to sound matched. This is why the edge-in is so important in smoothing transitions. You can have the level the ear expects and, by slowly increasing the level, restore it to normal.

An edge-in involves a small change in level but sometimes original material has far too much dynamic range. An obvious example might be classical symphonic music where the range from ppp to fff can easily exceed 50 dB. The traditional BBC guideline for dynamic range was a mere 26 dB. That is still valid for an item that someone is going to sit down and listen to. However, it is extraordinarily rare for anyone to do that. Radio broadcasting and audio presentations have the advantage that they can be absorbed by people who are busy, or on the move. In the modern noisy environment a dynamic range of 12 dB is about as wide as you can go without audibility suffering. Real life has a dynamic range of 120 dB, a power ratio of 1 000 000 000 000:1. How can we represent it with 12 dB, or only 16:1? The answer is by manipulating the ear's strength and weakness.

A fully modulated speech programme item will consistently peak to full level. Dynamic range is implied by cheating levels at transitions. A slow change of level is not perceived; a sudden change is. You can make an entry sound loud, be it a fortissimo passage or a shout, by dipping the level before the transition.

Say we have a presented item which is back-announcing a previous item on Mozart, followed by music played by The Who or The Rolling Stones. This is meant as a surprise. It is meant to sound loud. The presenter then voices over the music and must now sound louder than the music. Here, the answer is to creep down the level of the back announcement and start the music at full level. This is crept down and then dipped conventionally for the presenter's voice-over which is now at full level. This happy jig is kept up throughout the programme giving a sense of dynamic range yet is audible, even when heard in a car on a motorway.

11.3 Panning

There is a school of thought that says that all speech should be in mono in the centre and only music and effects heard in stereo. This may seem a little severe until you realize that all modern cinema films are balanced this way, as is much of the stereo sound on TV programmes.

In practice, speech inserts into programmes are in mono because they are easier to edit, if the image is not constantly moving around. A common convention is to put the presenter into the centre and then put contributors in panned 6 dB left or right. Panning more than this becomes gimmicky and potentially reduces compatibility if heard in mono. Another convention is to aim for a symmetrical balance with interviews panned equally left and right, or discussions panned symmetrically across the sound stage. (Studio interviews, using the presenter, should be recorded contributor panned hard left and the presenter panned hard right, using separate microphones. This gives total flexibility of positioning in production.)

11.4 Crossfading

Crossfades have several functions:

- equal power crossfades to hide the join between separate, but similar, sections of audio
- long passage of time/substantial change of location where a dip in level, combined with a change of sound, indicates a passage of time. This is effectively a modern contracted form of the old fade-out, pause, fade-in convention
- simple change of sound without level dip. This indicates a short passage of time and a small change of location (say, to get from one room to another).

11.5 Stereo/binaural

Conventionally, audio is heard on loudspeakers. Normally there are two of them but, for surround sound systems like Dolby stereo, there will be more. These surround systems either use special coder/decoders to play phase tricks with two channel sound, or they use multiple channels. DVD has standardized on a 5.1 channel surround system. The '.1' is an engineer's jokey way of saying that the sixth channel is low bandwidth, suitable only for conveying low frequency audio (earthquakes, spaceships, etc.).

Binaural sound has been around since the beginning of stereo when Clément Adèr relayed the Paris Opera to an exhibition in 1881. Visitors to the Paris Exhibition listened using two earpieces fed separately from different microphones at the theatre.

Techniques for recording binaural sound usually involve some form of dummy head. This can range from an actual model of a human head, with microphones in its ears, to something more stylized but just as effective; a Perspex disc approximately the size of a head, with small omnidirectional microphones mounted either side about 9 inches (22 cm) apart (the relative transparency of the Perspex disc makes it acceptable for slinging above orchestras at public concerts. A disembodied head might disturb the audience). There are also microphones designed to fit in a person's ears and thus a real head can be used.

Binaural seeks to emulate how we actually hear. The resulting sound is surprisingly good heard on loudspeakers and can be dramatically effective heard on headphones (open headphones work better than enclosed ones). When it works for you, a totally three-dimensional sound is heard. However, it is said that only 60 per cent of the population can get the best out of the system. A large proportion of the rest get a 3D effect but not one emulating real life. A common problem is everything seeming to be coming from behind the listener.

It works using phase and frequency response variations caused by the obstruction represented by the dummy head. This gives sounds outside the listener's head. For example, a mono voice panned fully left, heard on headphones, will seem to be coming from the left ear itself. A binaural sound coming from the left can seem to be metres further out. The phase accuracy and excellent frequency response of digital recording and editing make this an ideal medium for binaural sound.

Binaural recording can make an exciting alternative for productions aimed at headphone listeners, such as a headphone tour around a historic building. Although the phase accuracy of compact cassette is not that good, the binaural effect survives, provided everything is kept digital up to the master that the cassettes are copied from. Portable CD or Minidisc players preserve the directional effects because of their inherent phase accuracy. (Small timing errors between tracks are often expressed as phase differences. This, a difference of 1/48 000th of a second, is equivalent to a quarter of a wavelength at 1 kHz.

Because of the mathematical nature of waves, a whole cycle of audio is said to have gone through 360°. Thus, half way through a cycle is 180°. Here complete cancellation will occur if the two sources are mixed at equal level. Cassette machines have a degree of jitter where the track relationships vary; how much is dependent on how good the tape transport is.)

Mono sound, added to the mix, remains firmly 'between the ears' and this can be used to distinguish narration from other voices. There have been a number of dramas where characters, mute from a stroke, have had 'think speech' coming from within the listener's head, while all the 'real world' action takes place outside of the head.

11.6 Multi-layer mixing

Dramatized features sometimes need to create a sound for which there is no recording, such as the Biblical story of the Fall of Jericho. Here, building up the sound will require a great number of tracks. A large army can be simulated by sound FX recordings of suitably selected crowds, added on top of each other. If necessary, the same crowd can have different sections mixed with itself to make it sound larger. Various screams and shouts have to be added and panned appropriately, plus the trumpets that bring down the walls.

Before the arrival of digital multitrack editors this sort of thing was difficult to build up, because of the limited number of play-in machines, and often required multi-generation dubs, not to mention enormous tape loops precariously run round the room.

Such an 'epic' can be made separately and inserted into your program, either by mixing it down and inserting it as a stereo wave file or, if you wish, totally to integrate it with what precedes and follows it by appending its session (File/Append session) to your main programme session. The Append process will put the tracks on new tracks at the beginning of your programme session (to keep the Pan and Volume settings for the tracks). However, these tracks can then be grouped together so that they can be moved as a single item within the production.

12

Audio design

12.1 Dangers

Audio editors come with a multitude of 'toys' that you may never need – until that day when they rescue you from certain disaster!

Remember that, just because a toy is available, you do not HAVE to use it. There is no substitute for clear, interesting, relevant speech, or good music, well performed. With documentary, have faith in the power of the spoken word. One of the more depressing of my activities as a studio manager was working with producers who did not. They would ruin exciting speech by blurring it with reverberation or obscuring it with sound effects.

All the facilities available allow you, very easily, to ruin recordings. Audio and music technology is full of controls that, set in one direction, do nothing and, set in the other, make everything sound dreadful. Somewhere in between, if you are lucky, is a setting where something magical happens and a real improvement is obtained. When trying to find this setting, always refresh your ears regularly by checking the sound of the recording with nothing done to it.

12.2 Normalization

The first of these toys is normalization, which is so massively useful that you are likely to use it constantly.

One of the very first things that is done when making a recording is to 'take level'. The recording machine is adjusted to give an adequate recording level with 'a little in hand' to guard against unexpected increases in volume. However, that little bit in hand will vary, as will the volume of the speaker or performer.

Most recordists know that people, in general, are 6 dB louder on a take than when they give level – except those occasions when they are quieter! The result is that, even for those most meticulous with levels, different inserts into a programme will vary in level and loudness. While level and loudness are connected they are not the same thing.

The three illustrations (Figures 12.1–12.3) show the same piece of audio but at different loudness. Figure 12.1 illustrates audio which is at a lower level (6 dB) than the other two. Figures 12.2 and 12.3 are the same level but Figure 12.3 sounds much louder.

When we talk about level we actually mean the 'peak' level – we talk about audio 'peaking' on the meter. On average speech or music most of the time the instantaneous level is much less.

Audio, within the computer, is stored as a series of numbers so it is very easy for the computer to multiply those numbers, by a factor, so that the highest number in your audio signal is set to the highest value that can be stored. If this is done to all your audio inserts when you start, this will ensure that your levels are consistent.

However, their relative loudness may not be the same. It all depends on what proportion of your recording is at the higher levels. As a generalization, while the peak level is represented by the highest points reached on the display, the loudness is equivalent to the area of the audio display.

Figure 12.3, although showing exactly the same peak level as Figure 12.2, is much louder because it has been compressed.

12.3 Compression/limiting

Audio compression varies enormously in what it can do. It is not a panacea, as it can cause audibility

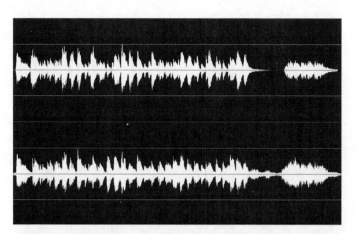

Figure 12.1 Audio example at –6 dB

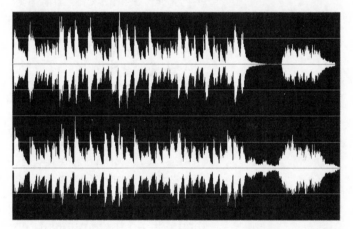

Figure 12.2 Audio example normalized

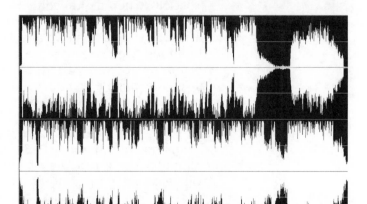

Figure 12.3 Audio example compressed

problems. Essentially, a compressor is an audio device that changes its amplification, and hence the level, from instant to instant.

You have control over how fast peak level is reduced (attack time). You have separate control over how fast the level is restored after the peak (recovery time). You also have control of how much amplification is applied before compression, by how much the level is controlled, and at what level the compression starts.

The simplest compressor that most people meet is the 'automatic volume control' (AVC) used by recording devices ranging from Minidisc recorders to telephone answering machines. A professional recorder should give the user the option of bypassing the AVC, to enable manual control of the level.

The automatic volume control has both a slow attack and a slow recovery time. Often it has just two settings – speech and music – with the music setting having even slower attack and recovery times. Even so, modern devices found on such things as Minidisc recorders are often relatively unobtrusive.

The normal advice, with analogue recording, is not to use AVC where the resulting recording is going to be edited. This is because the background noise goes up and down at the same time and any sudden change at an edit will be very obvious. With digital recording and editing the arguments are more evenly balanced, especially when digital editing is being used.

Analogue recorders overload 'gracefully' and the increase in distortion is relatively acceptable, for short durations. Digital recorders are not so forgiving. They record audio as a string of numbers and have a maximum value of number that they can record. This corresponds to peak level on the meter.

The compact disc standard, also used by Minidisc and DAT, is 16 bit, which allows 65 536 discrete levels occupying the numbers +32 768 to –32 768. Any attempt to record audio of a higher level than this will cause a string of numbers set at the same peak value. When played back, this produces very nasty clicks, thumps or grating distortion.

When recording items like vox pops, by the very nature of an in-the-street interview, there is likely to be a very high background noise. This has always been a classic case where switching off any AVC was regarded as essential, because of the resulting traffic noise bumps on edits. However, the likelihood of short- or medium-term overload is quite considerable. On the other hand, sudden short, very high level sounds such as exhaust backfires and gunshots can 'duck' the AVC to inaudibility for several seconds.

Digital editors make matching levels at edits the work of seconds. The danger of overload is such that this correction may be preferable to risking losing a recording owing to digital overload. In practice, it is down to the user to decide how good their recording machine's AVC is compared with how well it copes with overloads. They may prefer deliberately to under-record, giving 'headroom'.

Most portable recorders have some form of overload limiting which may allow you to return to base with 'usable' recordings, but it is better to return with good recordings. In the end, a saving grace is that it is the nature of these environments that the background noise is quite high and will drown any extra noise from deliberately recording with a high headroom.

Compression may make the audio louder, but it also brings up the background noise rather more than the foreground. This can mean that noise that was not a problem, becomes one. It will also bring up the reflections that are the natural reverberation of a room. This can change an open, but OK, piece of actuality into something that sounds as if the microphone had been lying on the floor pointing the wrong way.

If you are creating material for broadcast use you should also be aware that virtually all radio stations compress their output at the feed to the transmitter. While the transmitter processors used are quite sophisticated, they will be adding compression to any you have used. This can mean that something that sounds only just acceptable on your office headphones may sound like an acoustic slum on air.

There are standard options on any compressor that are emulated in any compression software. The compressor first amplifies the sound and then dynamically reduces the level (compression software is often found in the menu labelled as 'Dynamics'). Compressors can be thought of as 'electronic faders' that are controlled by the level of the audio at their input. With no input or very quiet inputs they have a fixed amplification (gain), usually controlled by separate input and output gain controls. Once the input level reaches a threshold level, their amplification reduces as the input is further increased. The output level still increases, but only as a set fraction of the increase in input level. This fraction is called the compression ratio (Figures 12.4 and 12.5).

A compression ratio of 2:1 means that once the audio is above the threshold level then the output increases by 1 dB for every 2 dB increase in input level.

A ratio of 3:1 means that once the audio is above the threshold level then the output increases by 1 dB for every 3 dB increase in input.

Compression ratios of 10:1, or greater, give very little change of output for large changes of input. These settings are described as limiting settings and the device is said to be acting as a limiter.

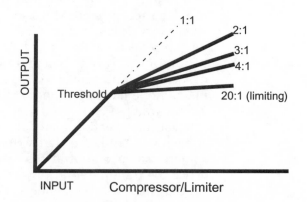

Figure 12.4 Conventional graph showing how the output audio of a compressor varies depending on the level of the input audio

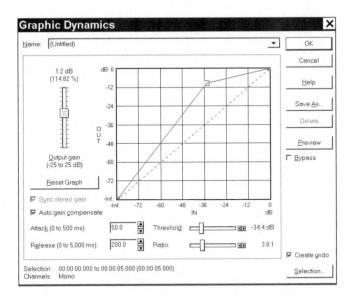

Figure 12.5 Compressor dialog from the Sound Forge linear editor allowing you to draw the graph of what you want

Hard limiting

Hard limiting can be useful to get better level out of a recording that is very 'peaky'. Some people have quiet voices with some syllables being unexpectedly loud. Hard limiting can chop these peaks off with little audible effect on the final recording, except that it is now louder. Cool Edit Pro has a separate transform for this.

Threshold

The Threshold control adjusts at what level the compression or limiting starts to take place. By the very nature of limiting, the threshold should be set near to the peak level required. With compression, it should be set lower; 8 dB below peak is a good starting place.

Attack

Control for adjusting how quickly the gain is reduced when a high level signal is encountered. If this is set to be very short then the natural character of sounds is softened and all the 'edge' is taken out of them. It can be counterproductive as this tends to make things sound quieter. Set too slow an attack time and the compression is ineffective. A good place to start when rehearsing the effect of compression is 50 ms. With musical instruments much of their character is perceived through their starting transients. Too much compression with a poor choice of attack time can make them sound wrong. On the other hand, a compressor set with a relatively slow attack time can actually improve the sound of a poor bass drum by giving it an artificial attack that the original soggy sound lacks.

Release/recovery

These are alternative names for the control that adjusts how fast the gain is restored once the signal is reduced. Set too short and the gain recovers between syllables giving an audible 'pumping' sound which is usually not wanted. This makes speech very breathy. A good starting point is 500 ms to 1 s.

Input

Adjusts the amplification before the input to the compressor/limiter part of the circuit. With software graphical display controls this may, at first, appear not to be present but is represented by the slope of the initial part of the graph.

Output

Adjusts the amplification after the compressor/limiter part of the circuit. Because it is after compression has taken place it also appears to change the threshold level on the output. Software compressors often have the option automatically to compensate for any output level reduction due to the dynamic gain reduction, by resetting the output gain.

In/out, bypass

A quick way of taking a device out of the circuit for comparison purposes.

Link/stereo

Two channels are linked together for stereo so that they always have the same gain despite any differences in levels between the left and right channels. If this is not done then you may get violent image swinging on the centre sounds such as vocals.

12.4 Expanders and gates

Cool Edit's dynamics can also act as expanders and gates. These work in a very similar way to compressor/limiters but have the reverse effect. With high level inputs they have a fixed maximum gain. Once the input level decreases to a threshold level then their gain decreases as the input is further decreased. When used as expanders the output levels still decrease but only as a set fraction of the decrease in input level. This fraction is called the expansion ratio.

A expansion ratio of 2:1 means that once the audio is below the threshold level then the output decreases by 2 dB for every 1 dB decrease in input level. A ratio of 3:1 means that the output decreases by 3 dB for every 1 dB decrease in input. Expansion ratios of 10:1 or greater give a very large change of output for little changes of input and the device is said to be acting as a gate.

Unlike compressors, which were originally developed for engineering purposes, the expander is an artistic device and must be adjusted by ear. Changing its various parameters can dramatically change the sound of the sources. Its obvious use is for reducing the amount of audible spill. This is usually restricted by the fact that too much gating or expansion will change the sound of the primary instrument. However, this very fact has led to gates and expanders being used deliberately to modify the sounds of instruments – particularly drums. Their use is almost entirely with multitrack music; bass drums are almost invariably gated.

Expanders and gates are bad news for speech. They change the background noise. The ear tends to latch on to this rather than listening to what is said. While I can imagine circumstances where they might help a speech recording, I have not yet met any.

12.5 Reverberation and echo

Reverberation (often abbreviated to 'reverb') and echo are similar effects.

Echo is a simpler form where each reflection can be distinctly heard (Echo ... Echo ... Echo ... Echo).

Reverberation has so many different reflections that it is heard as a continuous sound. For reasons more owing to tradition than logic, broadcasters often refer to reverberation as 'echo' and use the term 'flutter echo' for echo itself.

Reverberation time

The amount of reverberation is usually quoted in seconds of reverberation time. This is defined as the time it takes for a pulse of sound (such as a gun shot) to reduce (or decay) by 60 dB. This ties in well with how the ear perceives the length of the reverberation.

Devices

Reverb is the first and oldest of studio special effects. Originally it was created by feeding a loudspeaker in a bare room and picking up the sound with a microphone (Figure 12.6). Although this worked, there was no way of changing the style of reverberation. It was also prone to extraneous noises ranging from traffic, hammering, underground trains or even telephones ringing inside the room.

The next development was the 'echo plate'. A large sheet of metal – the plate – about 2 m × 1 m was

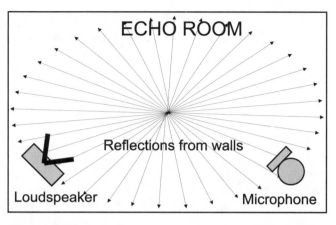

Figure 12.6

suspended in a box. There was a device to vibrate the plate with audio and two pick-ups (left and right for stereo) to receive the reverberated sound.

This worked on a similar principle to a theatre 'thunder sheet'. The trick was to prevent it sounding like a thunder sheet! This was achieved by surrounding it with damping sheets. By moving these dampers away from the sheet, you would increase the reverberation time (usually at the cost of a more metallic sound). Often, a motor was fitted allowing remote control. In a large building with many studios, a handful of the devices could be shared using a central switching system.

Digital

Studios these days use digital reverberation devices. The software used on these machines is equally able to be written for use on a PC. Like the separate devices it emulates, the PC will usually offer a number of different 'programs' with different acoustics (Figures 12.7 and 12.8). This will often include emulation of the old plates and room, but also include concert hall acoustics, along with special effects that could not exist in 'real life'.

Reverberation is a complex area as it triggers various complex subconscious cues in the brain. There are two basic pieces of information we get from how 'echoey'

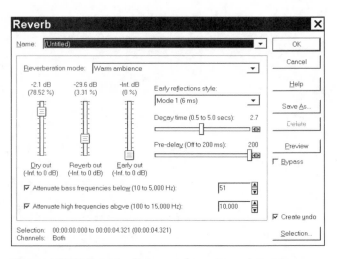

Figure 12.7 Reverb dialogue from Sound Forge

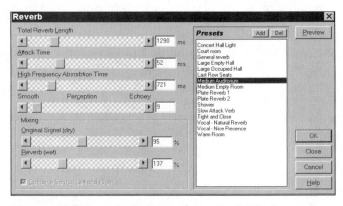

Figure 12.8 Reverb dialogue from Cool Edit Pro

a recording of someone speaking, or talking, is. They both interact, but need separately to be controlled. These are the size of the room, and how far away from the microphone the voice is. Move the voice further from the mic and the more echoey the recording becomes. If the microphone is held and the person walks into a larger room, the more echoey it becomes. However, we are also adept at distinguishing between different types of room. We can hear the difference between an underground car park and a concert hall, a hallway from a living room. How do we do this?

Let us imagine we are recording in the centre of a large, open acoustic, suspended in the air, as well as centred over the floor. Now make a sharp impulsive noise like a single hand clap, while making a recording. When we examine the waveform of the recording we will see that there is a moment of silence between the hand clap and the start of the reverberation. This is because reverberation is caused by sound reflecting off surfaces such as walls, floors and ceilings. Sound does not travel instantaneously. Its actual speed varies with the temperature and humidity of the air (one of the reasons, the acoustic of a hall can change so dramatically once the audience arrives, even if the seats have been heavily upholstered to emulate the absorption of a person sitting in them). Sound travels at roughly 330 metres per second, or about 1 foot per second. This means that if the nearest reflecting surface is 50 feet away then you will not hear that first reflection for 100/1000th of a second (100 ms). It is the size of this delay that is our major cue for perceiving how large a room a recording was made in. An unfurnished living room may be as reverberant as a cathedral but it will have short delay on the reflections from the walls (10–20 ms) and so can never be confused with the church.

Returning safely to floor level, we can make another recording, with someone speaking and moving towards, and then away from, the microphone. The microphone picks up two elements of the voice (as opposed to the ambient background noise from the environment). These elements are the sound of the voice received directly and the sound received indirectly via reflections.

As the voice moves nearer to, and further from, the microphone, the amount of indirect sound does not really change in any practical way. However, the voice's direct level does change. It is very easy to imagine that it is the reverberation that is changing. This is because a recordist will be setting level on the voice. As it gets closer so the louder it will become. The recordist compensates by turning down the recording level and so apparently reduces the reverb.

Theoretically, every time the distance from the microphone is doubled, the level will drop by 6 dB.

This is often known as the inverse square law. While this is true for an infinitely small sound source in an infinitely large volume of air, this is not totally true for real sound sources. A person's voice, for example, comes not only from their mouth but from their whole chest area. Only the sibilants and mouth noises (like loose, false teeth and saliva) are small so they disappear much faster than the rest of the voice (one reason why mic placement can be so critical).

As a generalization, if the delays on the reflections remain the same, then the ear interprets relative changes in reverb level as movement of the sound source relative to the microphone. If the delays change then the ear interprets this as the mic moving with the voice into a different acoustic.

The reason why reflections die away is that only a fraction of the energy is returned each time. Hard stone surfaces reflect well; soft furnishings and carpets do not. Such things as good quality wallpaper on a hard surface will absorb high frequencies but reflect low frequencies as well as does the bare hard surface.

Another reverberation parameter is its smoothness. Lots of flat surfaces give a lumpy, hard quality of reverberation. A hall full of curved surfaces, heavily decorated with carvings, will give a much smoother sound. This does much to explain why many nineteenth-century concert halls sound so much better than their mid-twentieth-century replacements, built when clean, flat surfaces were fashionable. So much more is now known about the design of halls that there is little excuse for getting it wrong in the twenty-first century.

When creating an artificial acoustic, all these mistakes are yours to make! As always with audio processing, it is always very easy to make a nasty 'science fiction' sound, but so much more difficult to create something magical and beautiful. The transforms within all good digital audio editors are provided with presets. It saves time to use them after modifying them slightly. When you find a setting that you really like, you can save it as a new preset.

Summarizing, a simple minimalist external reverberation device would include options such as:

- *Predelay*. This controls how much the sound is delayed before the simulated reflections are heard. This normally varies from zero to 200 ms. As sound travels at approximately 1 foot per millisecond (1/1000th of a second), a predelay of 200 ms (1/5th second) will emulate a large room with the walls 100 feet (ca. 30 m) away (200 feet there and back).
- *Decay*. This controls the reverberation time. This is the apparent 'hardness' or 'softness' of a room. A room with bare shiny walls will have a long reverberation time. A room with carpets, curtains and furniture will have a short reverb time.
- *HF rolloff*. This is an extra parameter that shortens the reverb time for the high frequencies only, making the 'room' seem more absorbent and well furnished.
- *Input and output*. Adjusts the input or output level.
- *Mix*. Controls a mix between reverberation and the direct sound you are starting with. Sometimes, there are separate controls labelled 'dry' and 'wet'. As a dry acoustic is one without much reverberation, so pure reverberation is thought of as the wet signal.

Let us now take a closer look at how Cool Edit Pro implements this. Reverberation and echo comes under the submenu of delay effects and the choice is potentially overwhelming! (Figure 12.9).

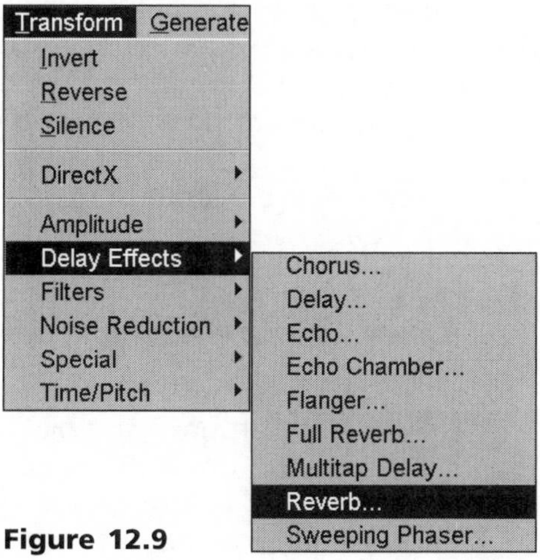

Figure 12.9

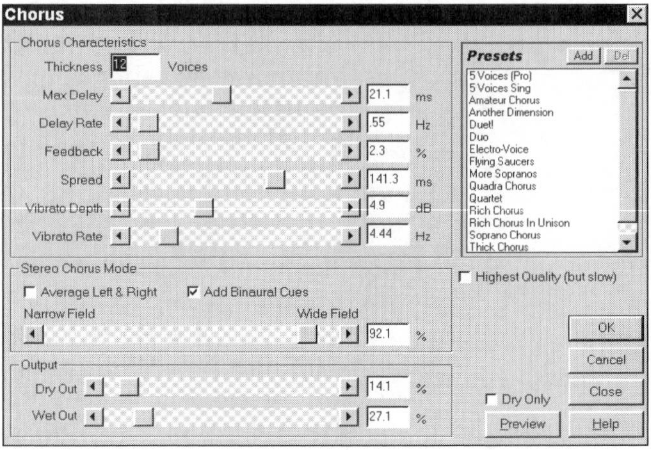

Figure 12.10

Chorus (Figure 12.10)

This is an effect used a great deal in pop music. By producing randomly delayed electronic copies of a track, it can make one singer sound like several or many. It can also be used to thicken musical tracks and other sounds. The random delays inherently cause a random pitch change. This can sound very nasty especially on spill behind a vocal.

Outside of music, chorus is not particularly useful except that the various options provided by chorus can produce interesting stylized acoustic effects, not dissimilar to adding echo. The variable delay pitch variation can help produce effects suitable for science fiction.

Using a recording of a single voice, give yourself a tour of the preset effects using the preview button. There may be something there you will find useful one day.

Delay (Figure 12.11)

This transform is the equivalent of tape flutter echo where the output of the playback head was fed back to the record input. While not the most subtle effect, it does have its uses. Again, give yourself a tour around the presets.

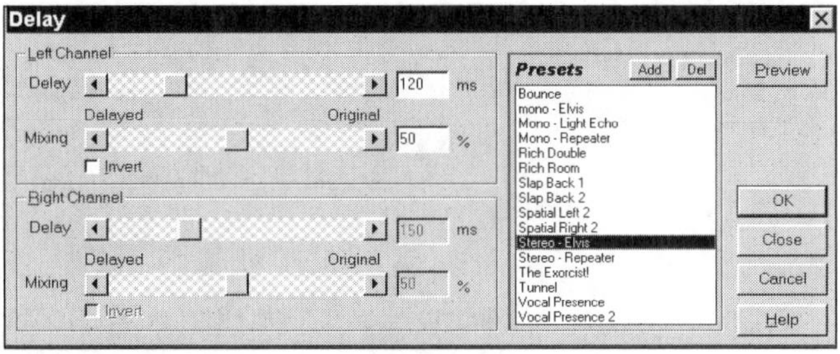

Figure 12.11

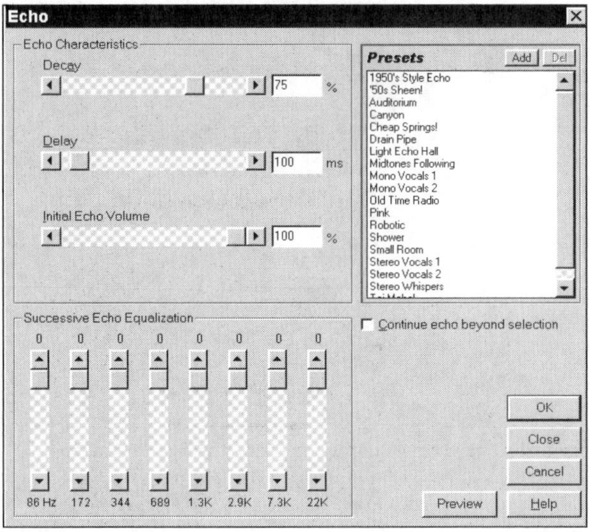

Figure 12.12

Echo (Figure 12.12)

This is very much like the previous transform with the additional option of adding equalization to the delayed echo. Again some useful presets are supplied.

Echo chamber (Figure 12.13)

This approaches designing a reverberant effect in a totally different way.

You are invited to enter the dimensions of the room you wish to simulate and where the microphone is

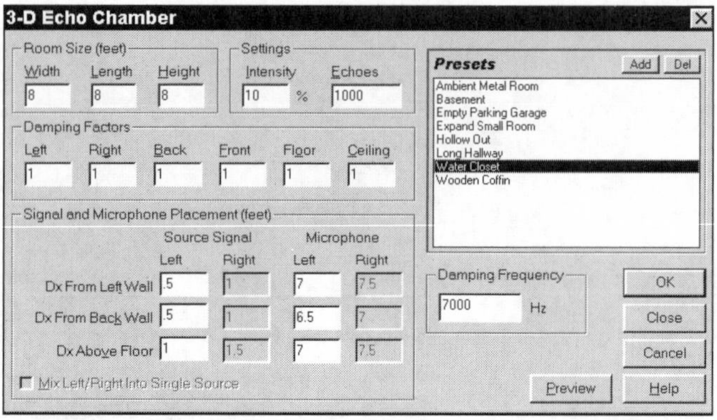

Figure 12.13

placed. You can also set the absorption (damping factor) of the floor, ceiling and each wall. Some useful dramatic environments are supplied as presets.

Flanger (Figure 12.14)

Legend has it that flanging was named by John Lennon of The Beatles. In the days before digital devices, they were experimenting in getting 'psyche-delic' sounds. They found that recording the same sound on two tape machines simultaneously would, when the two outputs from the separate playbacks heads were combined, provide a 'phasing' noise

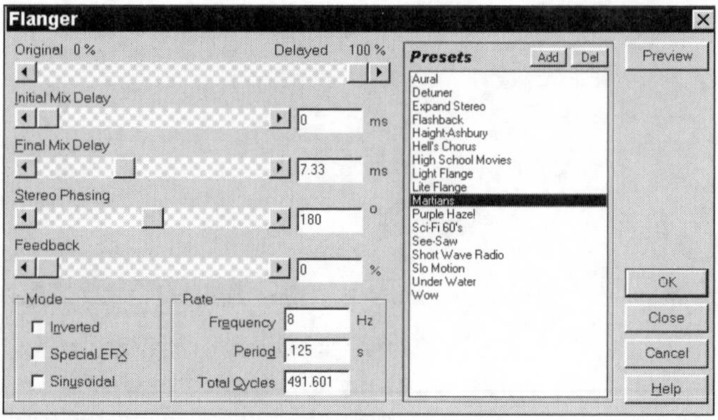

Figure 12.14

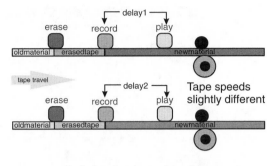

Original tape phasing

Figure 12.15

(Figure 12.15). This was due to tiny mechanical differences between the two machines giving a relative delay. This caused random frequencies to be cancelled out.

Flanging was so named because the cancelled frequencies could be changed dynamically by pressing a finger on the flange of the feed spool on one of the tape machines. The resulting 'skying' effect is irretrievably associated with late 1960s pop music. Both phasing and flanging are available and have their own useful presets.

Full reverb (Figures 12.16–12.18)

This is the all-singing, all-dancing reverb option. Because of its complication, I do not recommend using this unless you have a specific need and have the time and knowledge to navigate around the options. However, as before, there are some presets that you may find useful.

There are no less than three tabbed pages for settings (the vital ones controlling the level of the original signal (dry), early reflections and the reverb (wet) are repeated on each page).

The first page, 'general', contains the basic controls. Total length sets the reverberation time. Attack time sets the predelay (although Cool Edit Pro operates differently form some other reverb

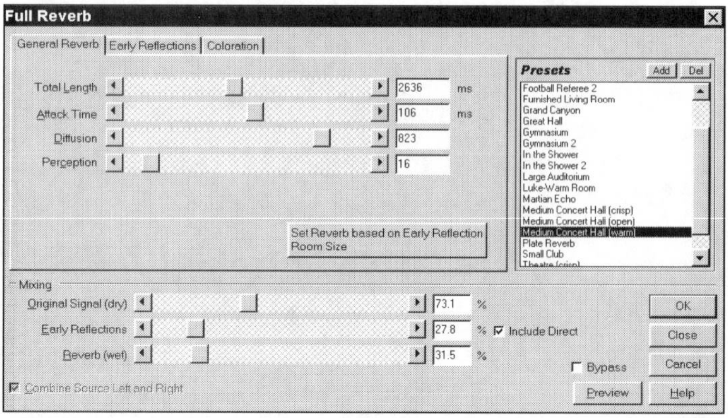

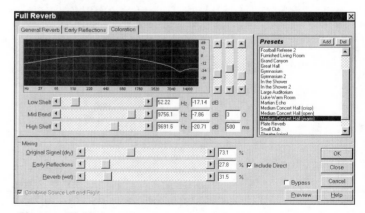

Figure 12.16

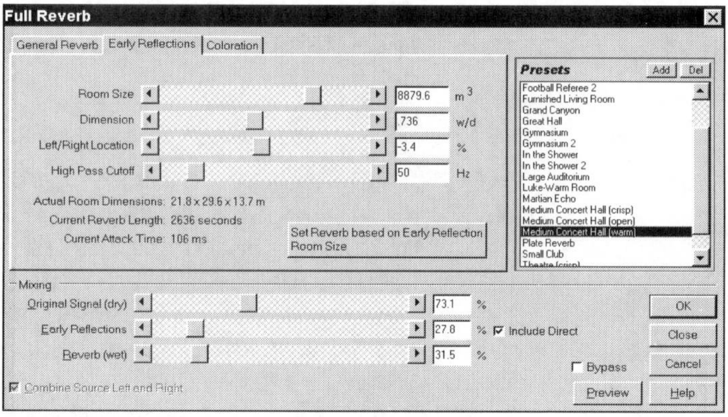

Figure 12.17

Figure 12.18

programs in this regard). Diffusion and Perception, between them, change the characteristic of the reverb from lumpy and echoey through to a smooth, continuous build-up and decay.

The second page allows you to control the early reflections, effectively controlling the room size.

The third page, 'coloration', gives you the ability to equalize the reverb within the algorithm. You can move the three bands, low, mid and high, around the frequency band. The amount of coloration is adjusted by the three sliders. As with the other transforms, the easiest approach is to use the presets and then modify as required.

Multitap delay (Figures 12.19 and 12.20)

This simulates the old tape loop delay systems. The effects produced are relatively primitive but suit certain types of music or stylized effects.

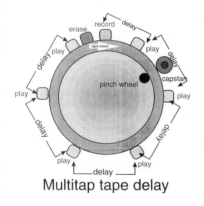

Multitap tape delay

Figure 12.19   **Figure 12.20**

Reverb (Figure 12.21)

This is the most useful general purpose reverberation transform. It is flexible, but still simple to operate. As well as a sensible range of presets, the reverb transform has the basic logical controls.

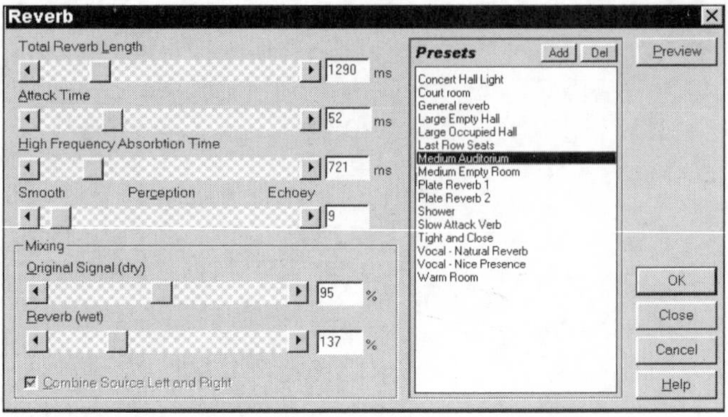

Figure 12.21

Total reverb length controls how long the reverb takes to die away.

High frequency absorption time gives you some control over 'how well furnished' the room is, while Perception gives control over the character of the reverb.

You can separately control the amount of the original and the amount of reverb in the final transform. This means that you can have the original sound mixed in while you are experimenting and then remove it to generate a reverb-only track for use in the multitrack editor.

This is the transform to use if you 'just want a bit of reverb' and you are not too critical as to subtlety.

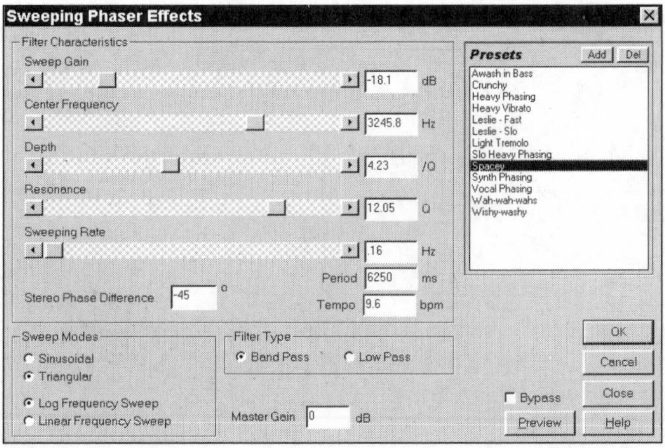

Figure 12.22

Sweeping phaser (Figure 12.22)

This is very similar to the flanger but with much more control over the effects available.

Again, start by using the presets and play with the settings to modify the sounds.

12.6 Equalization (Figure 12.23)

Cool Edit Pro has almost an embarrassment of equalization and filters. Equalization, or EQ as it is usually referred to, has two functions. The first is to enhance a good recording and the second, to rescue a bad one.

The best results are always obtained by first choosing the best microphone and placing it in the best place. This is not always possible and so, in the real world, EQ has to come to the rescue.

The most common need for speech EQ is to improve its intelligibility. A presence boost of 3–6 dB around 2.8 kHz usually makes all the difference. Here the simple graphic equalizer transform will be fine.

Another common problem is strong sibilance where 'S's are emphasized; reducing the high frequency

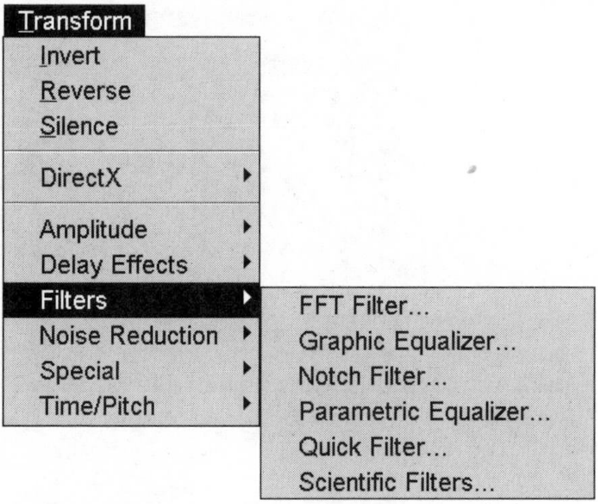

Figure 12.23

content can be advantageous (another option is to use the De-essing preset in the Amplitude/dynamics preset. This has the effect of reducing the high frequencies only when they are at high level).

Bass rumbles are usually best taken out by high-pass filters. This naming convention can be confusing, at first, as it has a negative logic. A filter that passes the high frequencies is not passing the low frequencies! Therefore a high-pass filter provides a bass cut, and a low-pass filter provides a high frequency cut. This sort of filter works without changing the tonal quality of the recording.

As always, it is best to avoid the problem in the first place. If your recording environment is known to have problems with bass rumbles, like the Central London radio studios which have underground trains running just below them, then use a microphone with a less good bass response. A moving coil mic will not reproduce the low bass well, unlike an electro-static capacitor microphone.

FFT filter (Figure 12.24)

FFT stands for Fast Fourier Transform. It is nothing less than the techie's dream of designing a totally bespoke

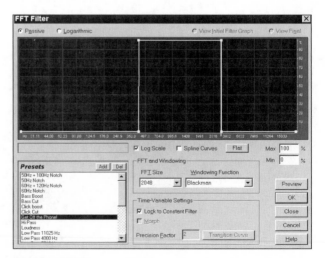

Figure 12.24

filter. Great fun if you know what you are doing. Totally confusing if you do not! Probably best avoided if you did not start your career as a technician.

Graphic equalizer (Figures 12.25 and 12.26)

This transform emulates the graphic equalizer often found on hi-fi amplifiers. Rather than just offering

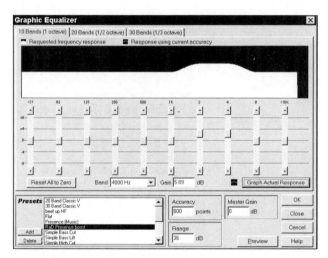

Figure 12.25

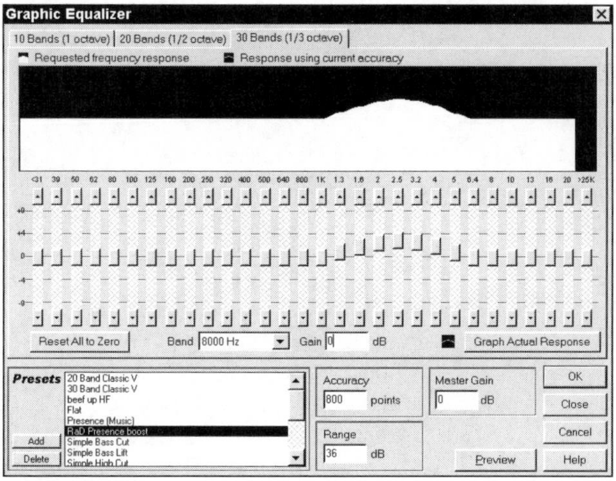

Figure 12.26

the five or six sliders of the average hi-fi, it gives you a choice of 10, 20 or 30 sliders. Each covers a single frequency band. The more sliders, the better the resolution. It represents a more friendly way of designing your own frequency response. It works rather better than analogue equivalents as it does not suffer from the analogue system's component tolerances. The illustrations show a setting for a typical presence boost using 10 or 30 sliders.

Parametric equalizer (Figure 12.27)

This is another techie-orientated transform.

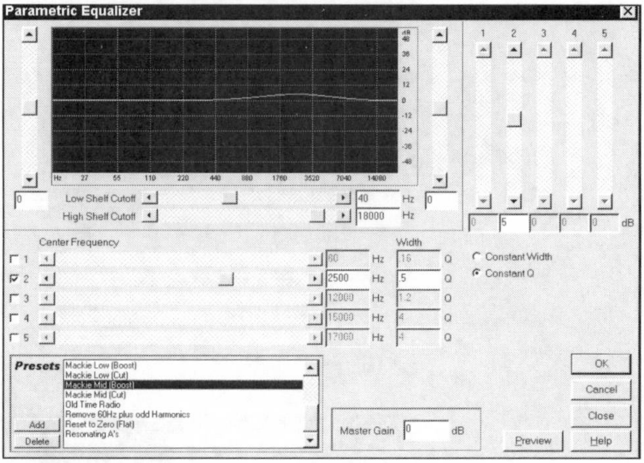

Figure 12.27

Notch filter (Figure 12.28)

This is a specialist transform which you will hope never to have to use. It allows you to set up a series of notches in the frequency response to remove discrete frequencies.

It comes preset with the most common disaster recovery task: getting rid of mains hum and buzz resulting from poorly installed equipment (theatre lighting rigs are a source of buzzes).

In Europe, the frequency of the mains power supply is 50 Hz. In the USA it is 60 Hz. Pure 50 Hz or

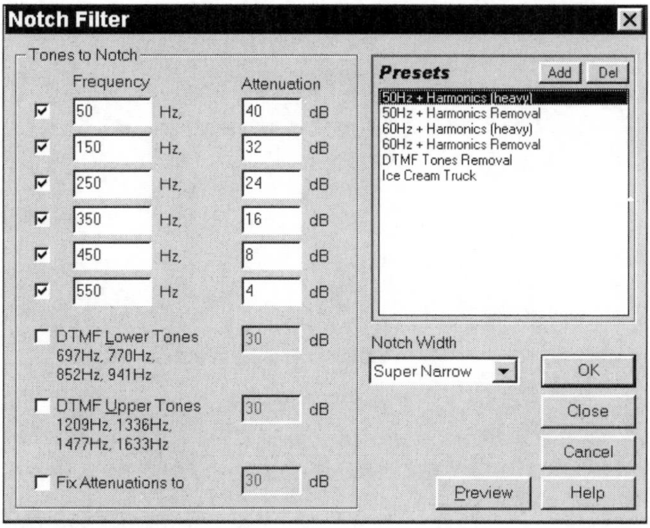

Figure 12.28

60 Hz is a deep bass note. However, mains hum is rarely pure. It comes with harmonics; multiples of the original frequency; 100 Hz, 150 Hz, 200 Hz, etc. (or their equivalents for a 60 Hz original). By 'notching' them out, the hope is that they can be removed without affecting the programme material too much. Except in severe cases this works surprisingly well.

The other preset options are to filter DTMF (Dual Tone Multi-Frequency) tones. These are the tones used to 'dial' telephone numbers. Some commercial radio stations use these to control remote equipment. Apparently they can end up getting mixed with programme material and have to be unscrambled from it.

Quick filter (Figure 12.29)

In many ways this is a little like another graphic equalizer except that the way it does things is slightly different. Its major difference is that it is two equalizers and implements a transition between the two settings. This can be useful to match different takes

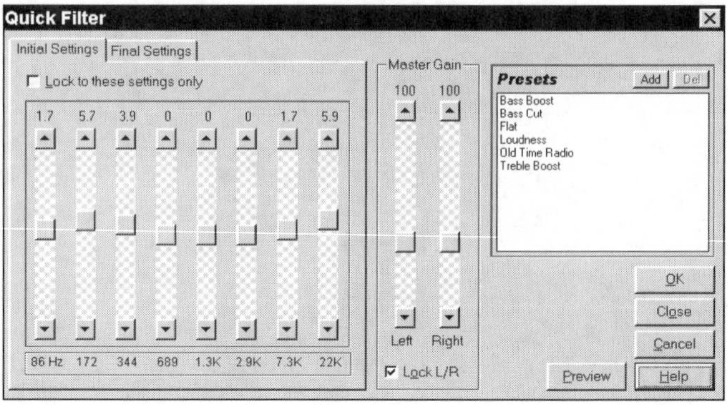

Figure 12.29

where not enough care has been taken to get them consistent in the first place.

It is also potentially useful for drama scene movements. An orchestra in the next room will sound muffled and then, as the characters move into that room, the high frequencies appear. This would have to be done in two stages with the Quick filter locked to apply the 'other room' treatment up to the point of the transition. It is then unlocked and the change made over a short section, butted onto the end of the treatment just applied.

However, I strongly recommend not doing it this way but to make a new track with the 'other room' treatment on it and use the multitrack editor to manage the change instead (see 'transitions' on page 107).

Scientific filters (Figure 12.30)

This is another build-your-own filter kit. Analogue filters are effectively made of building blocks, each of which can only change the frequency response by a maximum of 6 dB per octave. So, if you need something sharper, then an 18 dB/8ve filter needs three building blocks. This sort of filter is known as a third-order filter.

In the analogue domain, these filters are built from large quantities of separate components. Each of these components will have manufacturing tolerances

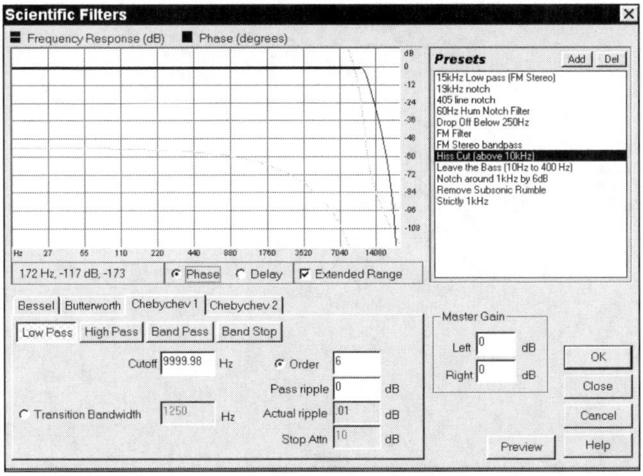

Figure 12.30

of 10 per cent, 5 per cent, 2 per cent or 1 per cent, depending on how much the designer is prepared to pay. This means that anything more than a third- or fourth-order filter is either horrendously expensive or impossible to make, because of the component tolerances blurring everything.

In the digital domain, these components exist as mathematical constructs. They have no manufacturing errors and therefore are exactly the right value. This means that a digital filter can have as many orders as you like, limited only by the resolution of the internal mathematics used.

Even so you cannot get something for nothing. Heavy filtering will do things to the audio other than change its frequency response. Differential delays will be introduced along with relative phase changes. Some frequencies will tend to ring. Leaving the jargon aside, what this means is that you may achieve the filtering effect that you want, but with the penalty of the result sounding foul.

The Scientific Filter Transform offers a set of standard filter options from the engineering canon. The graph shows not only the frequency response, but also the phase or delay penalties. In the end your ears will have to be the judge. If you have a tame audio engineer to hand – assuming that you are not one yourself – then you may persuade them to design you some useful filters optimized for your requirements and saved as presets.

12.7 Noise reduction

In the world of analogue recording, noise reduction refers to the various techniques, mainly associated with the name of Dr Ray Dolby, that reduce the noise inherent in the recording medium. In our digital world, such techniques are not needed except in some data reduction systems such as the NICAM system used for UK television stereo sound.

Noise reduction (NR), here, means techniques to take analogue recordings with their deficiencies and to 'digitally remaster' them. Noise reduction techniques can also be used to reduce natural noise from the environment, such as air conditioning or even distant traffic rumble.

Be warned that this is not magic; there will always be artefacts in the process. A decision always has to be made whether there is an overall improvement. Be very careful when you first use them as they are very seductive. It takes a little while to sensitize yourself to the artefacts and it is very easy to 'overcook' the treatment. As you would expect, the more noise you try to remove, the more you are prone to nasties in the background.

The most common noise reduction side-effect is a reduction of the apparent reverberation time as the bottom of reverb tails are lost with the noise. In critical cases, it can be necessary to add, well chosen, artificial reverb at very low level, to fill in these tails. Getting a good match between the original and the artificial requires good ears.

The basic method of noise reduction is very simple. You find a short section (about 1 second) of audio that is pure noise, in a pause, or the lead in to the programme material. This is analysed and used by the program to build up a filter and level guide. Individual bands of frequencies are analysed in the rest of the recording and, when any frequency falls below its threshold in the what is sometimes known as fingerprint, then this is reduced in level by the set amount. The quieter the noise floor the more effective this is.

However, trying to reduce high wide band hiss levels can leave you with audio suitable only for a science fiction effect. The availability of NR should not be used as an excuse for sloppiness in acquisition of recordings. Don't neglect obtaining the very best quality original recording in the first place.

The classic example of this is old 78 rpm recordings. These used a much wider groove. Theoretically, you could use an LP gramophone at 33 rpm to dub a 78 and then speed correct, and noise reduce, in the computer. This would sound foul. Played with the correct equipment, with purpose-designed cartridges and correct sized stylus, the quality from a 78 can be astonishing. The same applies to old LPs. Mono LPs had wider grooves than stereo records. A turntable equipped with an elliptical stylus will produce the best results from both mono and stereo LPs.

The most effective noise reduction can be achieved just by cleaning the record. Even playing the record through, before dubbing, can clean out a lot of dirt from the groove.

A really bad pressing, with very heavy crackles, will sometimes benefit from wet playing. Literally, pour a layer of distilled water over the playing surface and play it while still wet. This can improve less noisy LPs as well, but once an LP has been wet played, it usually sounds worse once dry again and needs always to be wet played. If it is not your record, then the owner may have an opinion on this!

The subject of gramophone records reminds us of the other major form of noise reduction, namely declicking. There are a wide number of different specialist programs that do this, as well as general noise reduction. They vary in how good they are, and how fast they are. If you are likely to be doing a lot of noise reduction then it may be worth your while to invest in one of these specialist programs.

Some of them, like the Sound Forge DirectX Noise Reduction plug-in, do just that, they plug into existing

Figure 12.31

Figure 12.32

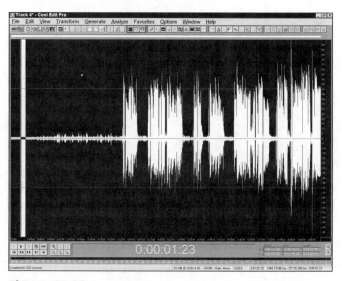

Figure 12.33

programs. Cool Edit Pro has its own effective, if a little slow, NR transforms but can equally well use a DirectX plug-in like the Sound Forge.

When Cool Edit Pro detects a DirectX plug-in it creates a new menu option under Transform which cascades sub-menus to allow selection of the plug-ins available (Figures 12.31 and 12.32) (these can include equalization and special effects as well). Some of these plug-ins can cost more than the program that they plug into!

Figure 12.33 shows the beginning of the vocal track from a cassette-based four-track machine with its dbx noise reduction switched off. Looking closely at the track you can see that some of the noise is from tape hiss and hum, but there is also spill from the vocalist's headphones. I have selected a couple of seconds where there is no spill, only machine noise.

I now need to obtain a noise print as a reference for treating the whole file. Both Cool Edit Pro and the Sound Forge plug-in have similar approaches. Cool Edit Pro also has a hiss reduction option. This produces a noise print looking solely for hiss (Figure 12.34).

However, if we take a noise print using the noise reduction transform, we can see that it has picked up the hum on the output of the four-track machine and raises the threshold at the hum frequencies

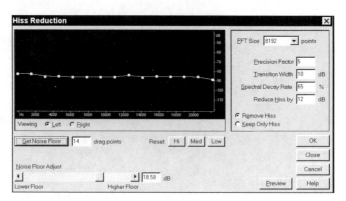

Figure 12.34

(Figure 12.35). The Sound Forge noise reduction plug-in does the same (Figure 12.36).

Having taken the noise print, we return to Cool Edit Pro and select the whole wave and initiate the noise reduction process. For clarity, in Figure 12.37, I used a high reduction setting available from the Sound Forge plug-in, so that you can clearly see

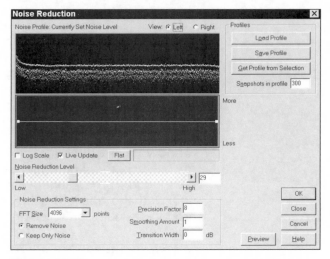

Figure 12.35

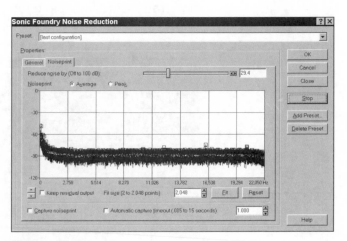

Figure 12.36

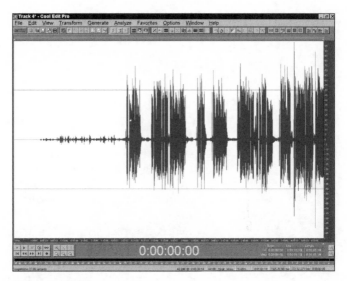

Figure 12.37

that it has removed the hiss but left the headphone spill. It will depend on the material whether this amount of noise reduction sounds acceptable. The spill can be removed selectively using the Silence transform.

Silence

This transform mutes the selected section of the wave file. Its main purpose is for music balancing, as you can use it to remove spill, coughs and sneezes from, say, a vocal track. This means that you do not have to set the volume envelope for each vocal entry and exit, but can guarantee absolute digital silence in breaks in the performance.

Don't confuse this with Generate/silence. The difference between the two is that the silence transform mutes a section of existing audio, making no change to its length. Generate/silence inserts a given number of seconds of silence at the cursor so that inserting 10 seconds of silence will make the file 10 seconds longer.

12.8 Electronic effects

Invert

The Invert menu option is a fairly specialized transform. All it does is to invert the waveform so that positive becomes negative and negative becomes positive.

In music balancing, this is the equivalent of the phase reverse switch on a mixing console. You can find, when you come to a mix down, that a microphone was out of phase with others used at the same time.

Microphones should be in phase. This means that their output voltages are going in the same direction; an impulse that makes the diaphragm move inwards on the microphones produces the same direction voltage. All the microphones should produce a positive voltage, or all the microphones should produce a negative voltage. It doesn't matter which, as long as they are the same. If they are out of phase, they tend to cancel, and the mix can sound a bit like a drainpipe.

The left and right channels of stereo should be in phase too as, otherwise, a central image seems to be coming from inside the head, rather than from between the speakers. Worse, if the recording is broadcast, people listening on mono receivers will not hear anything in the centre as it will cancel out.

Reverse

Reverse is a transform that does just literally that; it reverses the sound so that the beginning becomes the end. Its main uses are for producing reversed echo and alien voices.

Reversed echo is often used for magical effects: witches and aliens. It sometimes has a role in music, although it is an entirely unnatural effect.

A multistage process is used. First the programme material is reversed. Then it has a reverberation transform applied to it. That transformed recording is then reversed again. This restores the original speech, etc. to the right way round, but with the reverb now reversed. What was a reverb die-away now becomes a build-up. Satisfyingly weird effects can be created this way. However, the relative balance is critical so, again, it is best if the reversed reverb is created as a 'wet-only' track, so that it can be mixed with the original while you can actually hear the intended effect.

If you have a performer who is a good mimic, you can create a really good alien voice by getting them to record the lines and then reversing them. The performer now imitates the sound of the reversed speech – this takes some practice. He, or she, is then recorded imitating the backwards speech and then that recording itself is reversed. All being well, an inhuman voice will be heard speaking very strangely but intelligibly. This sort of trick has to be done in relatively short takes to work well.

12.9 Spatial effects

By spatial effects, we usually mean effects where the sound appears to be coming from somewhere other than between the speakers. A general atmosphere seems to fill the room, or a spaceship flies overhead, from behind you.

Techniques like Dolby stereo use extra loudspeakers, and special processing, to get their directional effect. DVD recordings can use five channels of audio to move sounds, but simple stereo can be beefed up, with a little simple manipulation.

Normally we think of stereo consisting of left and right channels. The relative volume of these channels controls where a sound comes from. A more sophisticated way of thinking about stereo is as an MS signal, MS standing for 'middle' and 'side' NOT mono and stereo. The middle signal consists of a simple mix of the left and right signals in phase. The side signal is again a simple mix of the left and right signals with the right channel phase reversed (inverted).

The middle signal is what is used as the mono signal in broadcasting. If you feed your stereo channel with a simple mono signal panned to the centre – say a presenter's voice – then there will be no side signal; the left and right channels are identical and, because the right-hand signal is inverted, they cancel themselves out.

As the mono signal is panned to one side the side signal increases until the sound is panned fully to one side. At this point, the middle and side signals are equal in level. Once you are thinking in this way, you can consider what happens when you increase the side signal so that it is larger than the middle signal. Doing this can pan the sound image outside the speakers. How effective this will be depends on the programme material. As the out-of-phase level is increased, the image will soon collapse and appear to come from inside the head.

With a really complex stereo signal, tweaking the side level can give dramatic dimensional effects. It is also a means of creating a fake stereo signal from a mono original.

Faking stereo

You sometimes need to fake a stereo sound from a mono source. This can range from a unique background sound to a period train, only ever recorded in mono, passing across the sound stage between the speakers.

There are several techniques open to the balancer. With continuous sounds such as applause, trains, atmospheres, etc., it is often possible to get a spread effect by playing two copies of the same (or different) sources. One is panned left and the other panned right with, perhaps, a third in the centre. If they get into close synchronization then compatibility problems may arise with phasing on a mono radio receiver. In Cool Edit Pro, this is easily done by using duplicate copies on different tracks and sliding the tracks relative to each other to get the right effect.

A better way, which gives no compatibility problems, is to play one source in mono and the other phase-reversed so that it only contributes to the stereo signal. This bonus sound is not heard by the mono listener as it cancels out.

Again duplicate tracks are used except that one (not both) of the duplicate tracks is converted to a stereo wave file and one channel (say the right) is phase-reversed (inverted). For a discrete sound, like an aircraft taking off or a train passing, sliding this track to delay the sound by 50 to 100 ms can produce good results, especially if the volume envelope on this difference channel is manipulated.

If you want a directional sound, like a train passing between the speakers, you need to do this in two stages. First generate a 'spread' train and mix this down to create a new stereo wave file. This can then be panned using the pan envelope control.

With atmospheres, very long delays can be used – try 30 seconds (if the FX is long enough). Changing the level of the out-of-phase signal in sympathy with the sound can improve realism.

Thunder claps can be particularly effective, if you run different mono claps on the in-phase and out-of-phase tracks. It can be difficult to find genuine stereo thunder that rolls around the room as effectively.

The same techniques can be used to beef up stereo sounds which do not come up to your expectations. Provided the sound is OK for any mono listener – if the recording is to be broadcast – then anything you add out of phase to your mix will be a bonus to the stereo sound without affecting the mono. If you can guarantee your recording will never be heard in mono then compatibility is not a problem. Remember that, in addition to mono radios, there are mono gramophones and mono cassette players.

12.10 Time/pitch (Figure 12.38)

Cool Edit Pro has two transforms to do with pitch changing and time stretching.

The first is the pitch bender. This is the direct equivalent of varying the speed of a recording tape. It can

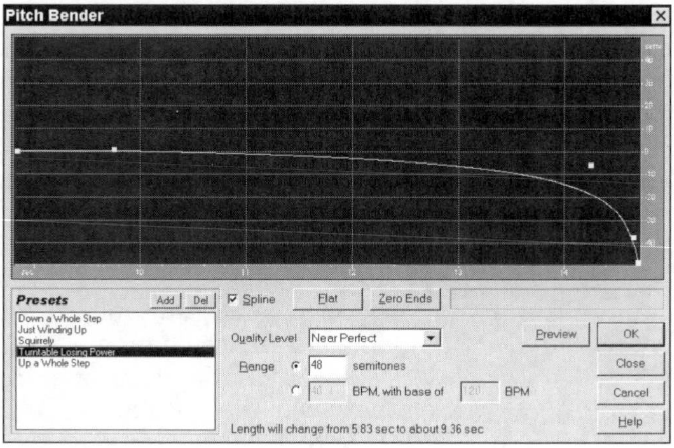

Figure 12.38

handle a fixed change, or make a transition from one 'speed' to another. You can draw the curve as to how it does this. As such it forms the ideal basis for a repair utility, for a recording made on an analogue machine with failing batteries.

What happens when the tape speed varies on an analogue tape is that, as the speed decreases, so the pitch decreases. If the tape is running at half speed, everything takes twice as long and the pitch is one octave lower. However, with computer manipulation this relationship no longer need be maintained.

Cool Edit Pro's next transform is Stretch (Figure 12.39). As well as more conventional manipulations,

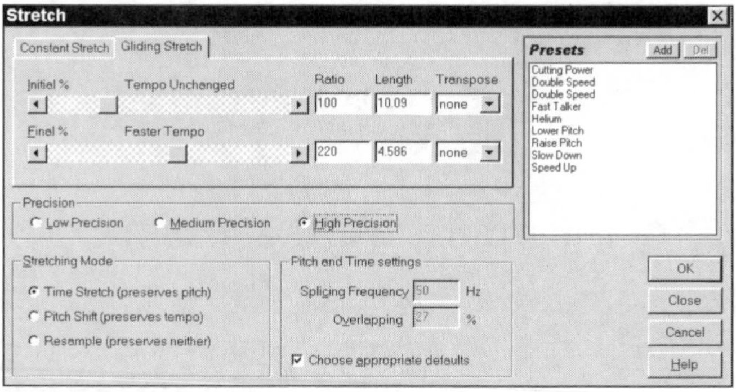

Figure 12.39

it allows you to change the speed of a recording without changing the pitch or, alternatively, change the pitch without changing the speed.

This is not done by magic, but by manipulation of the audio. Indeed there were analogue devices that did the same, just not very well. Even in the digital domain you are less likely to get good results with extreme settings.

Pitch manipulation is a specialized form of delay and works because, as in the chorus effect, changing a delay while listening to a sound changes its pitch while the delay is being changed. As soon as the delay stops being changed, the original pitch is restored whatever the final delay. Anyone who has varispeeded a tape recorder while it was recording and listened to its output will know the temporary pitch change effect lasts only while the speed is changing. It returns to normal the moment a new speed is stabilized.

A digital delay circuit can be thought of as memory arranged in a ring, as in Figure 12.40. One pointer rotates round the ring, writing the audio data. A second pointer rotates round the same ring, reading the data. The separation between the two represents the delay between the input and output. In a simple

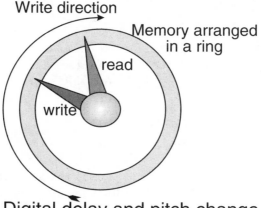

Digital delay and pitch change

Figure 12.40

delay, this separation is adjusted to obtain the required delay.

In a pitch changer, the pointers 'rotate' at different speeds. If the read pointer is faster than the write pointer, then the pitch is raised. If the read pointer travels slower than the write pointer then the pitch is lowered.

Quite obviously, there is a problem each time the pointers 'cross over'. This produces a 'glitch'. The skill of the software writer is to write processing software that disguises this. The software either creates, or loses, data to sustain the differently pitched audio output.

Uses

Time stretching. Skilfully implemented, you have a device that can take audio and time stretch it. A radio commercial lasting 31 seconds can be reduced to 30 seconds by speeding it up by the required amount. It can then be restored to the original pitch.

Singers, with insufficient breath control, can be made to appear to be singing long, sustained notes by slowing down the recording and pitch changing to the original.

Pitch changing. The pitch change can be useful for correcting singers who are out of tune.

Voice disguise. These devices are sometimes used to disguise voices in radio and television news programmes. There was also a fashion for them to be used for alien voices in science fiction programmes.

Music. With judicious use of feedback within the unit, pitch changing can be used to enrich sounds.

Special

The Transform special menu option offers convolution and distortion options which can be useful for producing science fiction distortions.

Weird effects for science fiction can be had aplenty, as virtually all the transforms, at extreme settings, produce very odd sounds which, hopefully, only happen when you want them!

12.11 Multitrack transforms

The Multitrack view of Cool Edit Pro has two transforms of its own. They are here because they combine, non-destructively, two different audio tracks to create a third. This is placed on a new track on the multitrack editor.

Vocoder (Figure 12.41)

A vocoder is a special effect which can make inanimate objects appear to speak or sing. If you want the leaves of a tree to talk or sing, then the vocoder may do what you want. It takes one sound (the process signal) and modulates it with the second (the control signal). Modulate a telephone bell with someone saying 'ring ring' and you could have a talking telephone bell.

To use Cool Edit Pro's vocoder you have to select two tracks and highlight the section to be processed. The easiest way is to have the control and process tracks adjacent to each other and to highlight both with one action. Place the mouse cursor over the first track, press and hold down the left mouse button at the beginning of, or just before, the wanted section

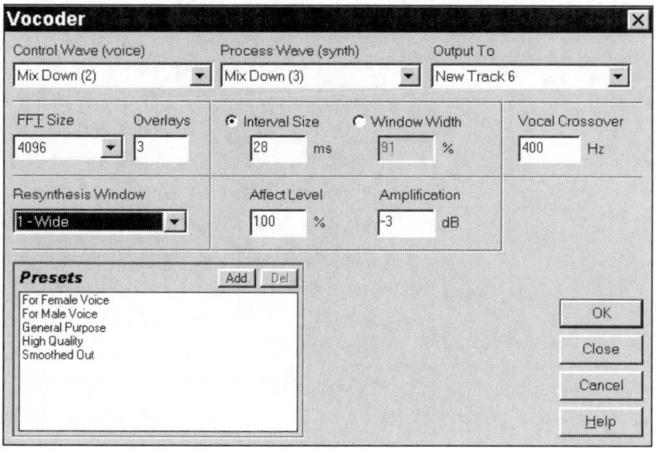

Figure 12.41

and drag the cursor down to the next track and along to the end of the section required. Both tracks will now be highlighted and the Transform/vocoder menu option will be enabled (this happens only if two tracks are selected).

If the two tracks are not adjacent, you can still do the same, except that the tracks in between will also be enabled. You need individually to deselect these by holding the Control key down and single clicking on each.

Selecting the section and then enabling two tracks by holding the control key down and single clicking on them will also work.

The top three boxes tell you about the two input tracks and where the transform will be placed. The left and centre drop-downs allow you to set which track is the control wave and which is the process file. The third box displays the next track empty at the selected area where the resulting transform will be placed.

Some presets are provided and the help file advises 'For ease of use, try checking "Window Width" and set it to about 90 per cent, use three or four overlays, a resynthesis window of 1 or 2, and FFT sizes from 2048 to 6400.' A fair amount of fiddling must be expected to get good results.

Envelope follower

The Envelope follower is selected in an identical way. Where the vocoder modulates one track by the other's audio, the envelope follower varies the output level of one waveform, based on the input level of the other. The amplitude map, or envelope, of one waveform (the analysis wave) is applied to the material of a second waveform (the process wave), which results in the second waveform taking on the amplitude characteristics of the first waveform. This lets you, for example, have a bass guitar line which only sounds when a drum is being hit. In this example you would have the drum waveform as the analysis wave, and the bass guitar waveform as the process wave.

Unfortunately, in version 1.2, there seems no way of inverting the envelope to provide a voice-over effect.

In addition to applying an amplitude envelope to a waveform, you can also alter the dynamic properties of the result, with a variety of settings to expand, gate, compress, or limit the resulting signal, which is why there is similarity between this dialog and the single waveform compressor transform dialog.

13

Reviewing material

13.1 Check there are no missed edits

In news and current affairs broadcasting, many compromises have to be made. Elsewhere, there is less excuse. Traditionally, with quarter inch tape, you would spool back from the end, with the tape against the heads, controlling the speed so that you could hear the speech rhythms. A missed edit, a gap or talkback, would show as a break in the rhythm. Any missed edits that were found could be razor-bladed in seconds.

With material prepared on an audio editor, life is not so simple. If the final item is transferred to CDR or DVD there is no way of listening to the recording while this is done. Transferring to DAT involves a real time copy which can be a good opportunity to review the item.

The snag is that if a missed edit is found then the copy has to be restarted from the beginning. Ironically, this can lead to a programme being prepared entirely using digital technology only to be transferred to analogue quarter inch tape so that any missed edits can be edited the old fashion way, with a razor blade, rather than having to start the transfer from the beginning.

13.2 Assessing levels

Very rapidly, the eye becomes used to assessing levels from the appearance of the waveform display.

However, there is no substitute for listening to the item in one go, on good equipment.

You should also make a habit of listening to your material in the environment that it will be used. If you are making a promotional cassette to be heard in a car, then listen to your tape in a car, preferably a noisy one so that you can check that levels don't drop to inaudibility.

If the item is to be used on the telephone, as an information line recording, then this presents a virtually unique balancing challenge. This is genuine monaural sound: one-eared. Check your balance with headphones worn so that only one ear is used. If you have facilities to feed the audio down a phone line, then do so. Remember, even if you have been asked to provide the material as analogue audio, it will almost certainly be reduced to, at best, 8-bit digital audio on the play-out device. Lower bit rates are often used, with the resulting quality being much below what the telephone system (which is an 8-bit system) is capable of.

13.3 Listening on full quality speakers

This will reveal every deficiency in your recording. It becomes a production decision as to what the listener will hear the recording on. It is as important to listen on poor speakers. Music studios always have a set of small speakers to check balances on low bandwidth speakers. A good balance travels well. If your item sounds fine on good speakers, and on a portable cassette player, then it is likely to be successful.

The 'grot' speaker also reveals whether effects that sound impressive on big speakers, or headphones, also work on average equipment. A classic mistake is to use a deep bass synthesized sound for a dramatic heartbeat. While sounding effective on big speakers with extended bass response, the same sound may sound like low level clicks on a transistor radio.

There is a definite difference between 'hi-fi' loudspeakers and 'monitoring' speakers (the speakers that come with PCs are neither). Hi-fi speakers are designed to make everything sound good and well balanced. Monitoring speakers are designed to be analytical. They are your equivalent of the doctor's stethoscope. You want to be able to hear things, and correct them, before your listeners hear them. This is why sound balancers monitor at higher levels than ordinary mortals. They wish to hear inside the balance. They will also 'dim' the speakers from time to time to check realistic levels on both their monitoring speakers and the 'grot' speakers.

14

Mastering

14.1 Line-up and transmission formats

The mastering process is concerned with producing the final product that is actually going to be used. This could be for use by yourself; say, a CD of audio illustrations for a talk, or it could be a multitrack tape for a Son et Lumière.

It could be a programme for broadcast. Here the final recording has to be prepared in the correct way to fit in with the requirements of the broadcaster as to levels and line-up information.

The requirement will differ between organizations and reflect their internal practice. For example, at the time of writing, BBC Network Radio requires programmes provided on compact disc to be recorded with a peak level of –4 dB. All you need to match this requirement is to normalize your programme to –4 dB and burn it to CD.

Increasingly broadcasters are accepting material on CD-ROM as a .WAV file. They usually require that the professional sampling rate of 48 kHz is used for these. Wave files can also contain formatted text information. Cool Edit Pro provides dialogs for entering this (Figures 14.1 and 14.2).

There is also a European Broadcasting Union standard for including text in the wave file to provide production information.

Line-up

Quarter inch tape and DAT usually need to have a line-up tone, or tones, at the front of the

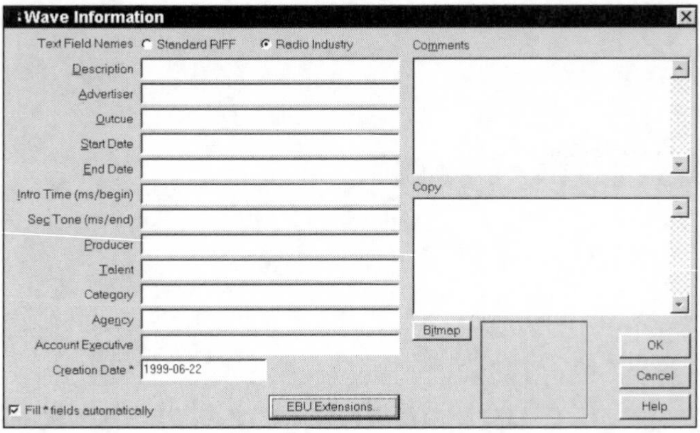

Figure 14.1

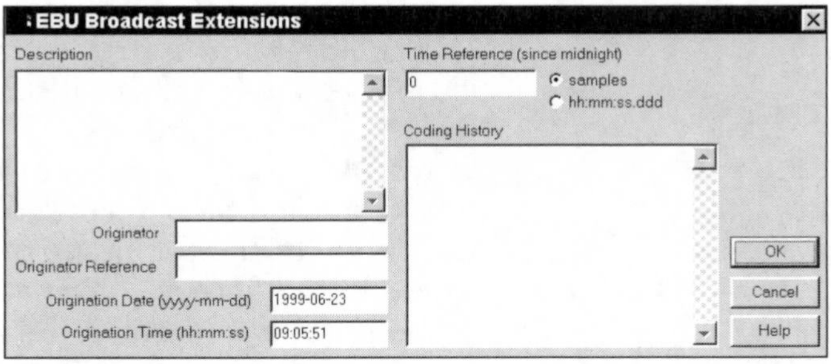

Figure 14.2

programme. Here BBC practice is for 20–30 seconds of, nominally, 1 kHz tone on both tracks recorded at 11 dB below peak programme level. Analogue tapes are also required to have a section of 10 kHz tone, so that head alignment errors can quickly be detected.

The 'Generate' menu title covers useful tools here.

14.2 Generate silence

This option does just that, it generates a set number of seconds of silence. It is different from the silence in the Transform section which mutes the highlighted section of the recording. Here, the silence is generated as a new recording – any audio recording after the cursor is pushed onwards to make way for the silence, i.e. rather than being replaced by the silence. This means that you can insert a silence between a line-up section and the programme material.

14.3 Generate noise

This is a specialized option. Generate noise means just that. You have a choice of three types of hiss: white, pink and brown.

How can noise have a colour? Well, the analogy is with the light spectrum. The three noises are hisses containing all the frequencies of the sound spectrum, just as white light contains all the colours of the rainbow. As we all know, there are different types of white: warm white, cold white, etc., so there are different colours of hiss.

White

White noise has equal proportions of all frequencies present; each discrete frequency has the same energy present as any other frequency. Because the human ear is more susceptible to high frequencies, white noise sounds very 'hissy'.

Pink

Pink noise has a spectral frequency similar to that found in nature. Each octave has the same energy as any other octave. As you go up the octave range, each octave has 'more frequencies' in it; so for every octave increase the energy for each discrete frequency is halved. If you look at the spectrum analysis of white noise it appears to have a flat frequency response but pink noise has one that falls 3 dB for every octave.

It is the most natural sounding of the noises. By equalizing the sounds you can generate rainfall, waterfalls, wind, rushing river, and other natural sounds.

Brown

Brown noise has much more low-end, and there are many more low frequency components to the noise.

This results in thunder- and waterfall-like sounds. Brown noise is so called because, mathematically, it behaves just like Brownian motion. This is how molecules move in a nice hot cup of tea!

As has been suggested, these three types of hiss can be dragooned into helping a sound effects mix. With a little top cut, brown noise can take on the role of 'city skyline'. This was the background rumble that, for decades, was the sound of silence on television. When they had mute film, a touch of skyline gave a sense of something going on. Legend had it that it was the sound of Victoria Falls played at half speed.

This sort of rumble added to a sound effects mix can give it a depth it might not otherwise have. Like a spice in cooking, it improves without being distinguishable.

Pink and white noise are also very useful test signals to check out audio systems, including loudspeakers. Resonances and frequency response dips show up quite clearly. The high frequency losses of recording systems such as analogue tape recorders are also easily checked. However, be warned: errors of a fraction of a decibel can be heard and this may well be within the maintenance tolerance of the machine.

14.4 Generate tones (Figure 14.3)

This can generate quite complex tonal mixes but, for our purposes, a simple pure sine wave tone at a specified level is all that is required. The illustration shows a setting for 900 Hz tone at 11 dB below peak which, at the time of writing, is the line-up level used by BBC Radio (PPM3¼ 1 kHz can also be used). The BBC also requires a 30 second section of 10 kHz tone at −18 dB below peak on analogue tapes. This line-up sequence should be put at the front of the programme on your computer so that, if you are dubbing your final programme to quarter inch, the whole sequence

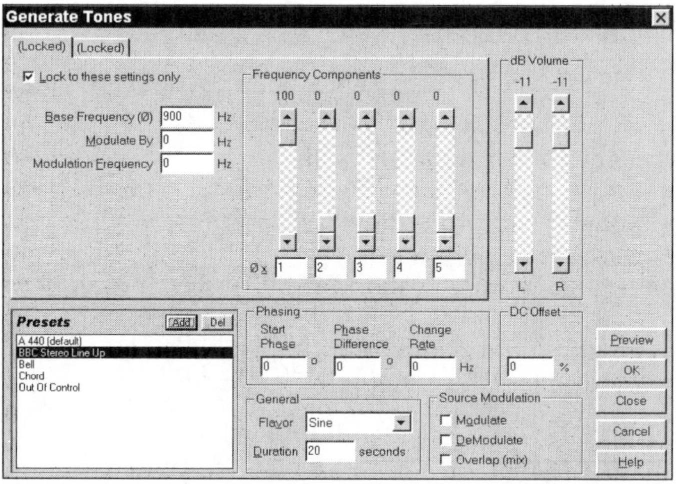

Figure 14.3

from line-up to programme end is dubbed in one pass.

14.5 Changeovers

If quarter inch tape is being used, then the maximum time available on a 10½" NAB reel is 30 minutes. Programmes longer than this have to be on more than one reel, and each reel has to be cut for a changeover in such a way that it can be performed perfectly without rehearsal. While this is a rapidly disappearing requirement it is still worth considering how best to cut a changeover.

Ideally this should take place at a change of item, preferably where there is a slight pause. (If the changeover is botched then the pause just gets slightly longer – with presented programmes it is regarded as good practice to end the first reel with the presenter's cue and start the second reel with the insert, or the music. The logic is that, if the changeover is missed, it does not sound as though the presenter has gone to sleep but is obvious that it was the technician!) The pause should always end the first reel NOT start the second.

14.6 Mastering process

The mastering stage is the last chance you have to check that the levels are consistent and that the overall recording is as loud as it can be. Picking off isolated peaks and reducing their level and then renormalizing can give a substantial extra level with no audible consequences.

DAT and analogue masters are created by the simple process of dubbing from the computer to the external recorder. Mastering CDs can be as simple, if an external CD recorder is used. However, these often can only use the more expensive consumer blanks, and share, with the other media, the need to copy in real time. CD writers for computers are relatively cheap and can burn CDs at 2, 4, 6, 8 or more times speed. To do this special software is required. This requires you to specify if you are burning an audio CD, a CD-ROM or a mixed format. Tracks are dragged into a window and the CD burnt (see Chapter 3 on 'Hardware and software requirements').

If you are planning to produce actual CDs then more sophisticated software is needed. Simple software works on the basis of one wave file per track. This need not necessarily be so. Software such as Sound Forge CD Architect can use one, or more, wave files and it is down to you how they are spaced and where the track markers are (Figure 14.4).

For CDs of events, you can vary the length of each inter-track gap or have none at all. You can even continue the audio through the inter-track gaps. For a CD of a concert you can arrange for each song to begin each track but for speech links and so on to be in the gaps between the tracks (the CD player counts down the seconds in the gap). The CD will play through continuously, but a track skip will take you directly to the next song.

If you have access to professional or semi-professional CD players which will make use of them, you can also add index marks.

A CD can have a maximum of 99 tracks but each track can have 99 index marks. The original idea was that a track would contain a single classical work,

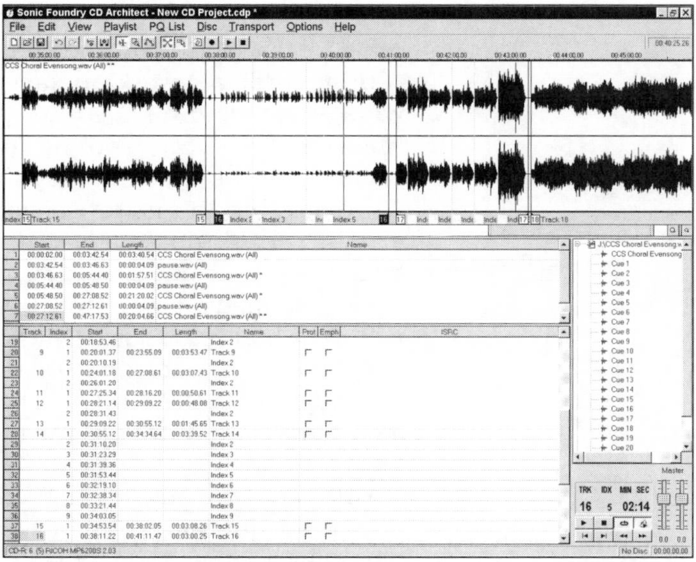

Figure 14.4 Sonic Foundry CD Architect CD burning screen

with each movement marked by index marks. These days just about the only CDs that have index marks are Sound FX CDs, as they have a multitude of short sounds which would soon use up the maximum of 99 tracks. So, for example, a track of 20 door slams would have each slam marked with an index mark.

Index marks can be useful to you for archival purposes, allowing you to mark individual items without starting a new track. Unlike tracks, index marks do not appear in the table of contents and the CD player literally has to scan through the disc to find them, so index searches are slightly slower than track searches.

14.7 CD labelling

Surprisingly, the most fragile surface of a CD is the label side, not the playing side. Scratches on either side can cause errors. Many CDRs come printed with labels for handwriting their contents. This is perfectly possible, but care must be taken to use a pen certified as suitable. A ball-point pen will just damage the CD and some ink solvents will dissolve the protective layer.

Using ordinary stick-on labels may, or may not, work, depending on the glue used on their backing. However, they can end up damaging your drive, rather than the CD, as it may be being spun at 40 or 50 times the standard speed and the off-centre weight can disrupt reading the disc and strain the bearings.

Purpose-made labels for CDs can be purchased. Different brands are usually associated with different proprietary applicators that will centre them properly. While these seem expensive, the applicators also come with software for designing and printing onto the labels.

You can also buy transparent labels to use with unprinted disc blanks to make the CDR label look more like a conventional CD. However, getting these onto your CD without air bubbles is more of an art than a science. Some people, who wish doubly to protect their CDs, put the transparent labels on top of the paper printed ones and also make them impervious to liquid spills which make the inkjet printing run. Ensure that the labels you use are suitable for your printer as inkjet and laser printers have different requirements.

14.8 Email and the World Wide Web

The requirements for the Internet are constantly changing but the constant for all but those with expensive lease lines is that there is a limit to the amount of data that can be sent every second. What is now often called POTS, Plain Old Telephone System, can at the very best with a V90 modem get 6–8 Kbytes per second of data, and even that is in one direction only – from the Internet Service Provider to the user. Upload speeds in the opposite direction are usually half that.

ISDN can offer those sorts of speeds reliably in both directions or twice that if both channels of an ISDN circuit are combined (usually at the penalty of being charged as two calls. While weary broadcasters are apt to translate ISDN as the 'It Sometimes Doesn't Network', it actually stands for 'Integrated Services Digital Network'. It is capable of handling any data, be it telephone, fax, graphics or even stereo music. The main complications are due to a plethora of standards, many of which are not, quite, compatible with each other). Lease lines offer virtually any speed that you want but at a price.

However well you are equipped, you want audio that you put onto the Internet to be accessible to users who are not so well equipped. This means that there is a need to make files as small as possible with a minimum loss of quality. A CD quality recording takes up 10 Mbytes for every minute. This makes it impossible to play in real time from the Internet as well as occupying a great deal of space on the hard drive of the server where it will be stored.

There are simple ways of reducing this but with quality penalties. Mono, rather than stereo, halves the size instantly, as does halving the sampling rate. Reducing the bit rate from 16 to 8 halves the size yet again. By the time you have done all this the quality is getting decidedly iffy.

What is needed is some way of compressing the audio data so that it occupies less space. Immediately ZIP files come to mind. They are almost universally used as a way of transferring other files. It makes the files smaller and they take less time to be sent. A ZIP file uncompresses its contents to exact copies of the originals. This works very well for text, but very badly for wave files. It can even happen that the zipped version of a wave file is actually bigger than the original.

What is needed is a different way of recording the data so that it sounds as good as possible, while throwing away as much data as possible. These are known as 'lossy' forms of compression, unlike the lossless ZIP file. Minidisc uses such a system, so that a 140 Mbyte Minidisc holds the same amount of audio as a CD containing 720 Mbytes of audio. This uses psycho-acoustic tricks so that it only records what we hear, not the background material that is hidden by the foreground.

The Microsoft ADPCM (Adaptive Differential Pulse Code Modulation) form of the wave file (it too has the extension .WAV) uses about a quarter of the data by sending information about the differences between samples rather than absolute values. A 1 minute stereo file that starts out at 10 336 Kbytes reduces to 2600 Kbytes. The quality stands up very well – it is mainly the high frequency transients that get subtly lost. Converting to mono and halving the sample rate would reduce that to a quarter, one-sixteenth of the original size.

Real Audio, and Real Media, files are designed to be sent to a modem. They can be 'streamed'. Their associated playback software, which can be integrated with Internet browsers, can play the files as they come in rather than having to wait for the entire file to be downloaded (there is a short pause while an audio buffer is filled). They can be optimized for the intended delivery modem. Our 10 Mbyte speech file can be delivered as a 118 Kbyte mono Real Audio file. The penalty is a phasiness that sounds as though the speech was recorded in an enclosed telephone box. A larger Real Media file of 367 Kbytes sounds better.

Cool Edit Pro can save in these formats but does not have a direct facility for loading them, as they are intended for delivery, not for editing. If you do need to transfer these files into the editor – maybe you are doing an item about the Internet – you can record the output of Real Audio/Media by setting Cool Edit Pro to record the output of the sound card handling the Windows output.

Another format is MPEG (Motion Picture Experts Group) which comes in various flavours with extensions of MP1, MP2 and MP3. MP3 is becoming commercially used as it is very efficient in providing low data, good quality sound in stereo.

Whatever your final format, it is especially vital that the sound file is properly prepared. Do everything at full 16-bit resolution and your normal sampling rate. Only when everything is ready do you reduce the sound to the lossy compression format.

You may wish to try adding a little more mid to high frequency presence to the sound to help it punch through. Use the hard limiter to lose isolated peaks. Maybe a little audio compression will help. Save the audio to your chosen format and listen to it critically. Repeat until you are satisfied.

BEWARE when you SAVE AS to, say, a Real Audio format, Cool Edit Pro's filename extension will change at the top of the screen, but its buffer will still contain full resolution, uncompressed audio. You have to close the file and reload to hear the result.

Cool Edit Pro can load and save a large number of different formats. Some are obsolete but still found. Others are used by other types of computer (such as AIFF for Apple Macintosh computers). The exact formats that you can load will combine those supported directly by Cool Edit, plus those supported by Windows and any added by your sound card software.

14.9 Analytical functions

Digital audio files consist of a large quantity of numbers. As such they can be analysed statistically and produce, at worst, pretty pictures and, at best, useful diagnostic information about an audio problem. This may be particularly valuable for older users who are losing their high frequency hearing and allow them to see problems that only others can hear.

Spectral view

Normally when using Cool Edit the waveform view is used. This shows how the amplitude of the signal varies with time. But there is an alternative view, the spectral view. This analyses the spectral content of the signal. Much of the time this is good for making a pretty picture on your screen and little else. However, it can be helpful in tracking down audio problems. Spectral view displays a waveform by its frequency components, where the height represents the frequency and, as usual, time is represented

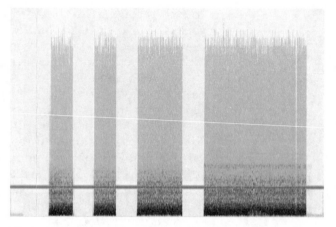

Figure 14.5 Spectral view revealing a low level 3 kHz tone

horizontally. This will show you which frequencies are most prevalent in your waveform. The greater a signal's amplitude component within a specific frequency range, the brighter the displayed colour will be. Colours range from dark blue (little audio in this range) to bright yellow (frequencies in this range are high in level).

As a diagnostic tool it will show you frequency response problems such as a suck-out at particular frequencies (perhaps resulting from a head alignment problem in an analogue original). It will also show, as in Figure 14.5, the presence of low-level tones and help you identify their frequency. Know the frequency, and you are half way to finding the source of the problem.

Frequency analysis (Figure 14.6)

The Frequency analysis window contains a graph of the frequencies at the insertion point (yellow arrow cursor) or at the centre of a selection. This window 'floats', meaning that you can click in the waveform on the main Cool Edit Pro window to update the analysis while the Frequency analysis window is on top.

The information in this dialog is like one 'slice' or line in the Spectral view of the waveform. The most

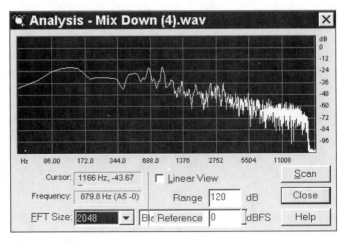

Figure 14.6

prominent frequency is interpolated and displayed in a window below. You can move the mouse over the graph area to display the frequency and amplitude components of that frequency. Where there is a single instrument playing, you can often see its harmonics quite clearly.

You can control how finely the frequencies are analysed. Provided this is not too rigorous for your computer, there will be enough processing power for the graph continuously to update itself while the file is being played. This dynamic display can often reveal low level, or high level, tones, etc., that cannot be spotted otherwise. For example, high frequency noise may be present where there is no audio activity. This should be filtered for the benefit of people who can hear it. Similarly, a recording made from FM radio may show signs of break-through from the 19 kHz tone that synchronizes the stereo system.

You can also generate a step-by-step animation by clicking on the main waveform window and then holding down on the right arrow key. As the cursor scrolls across the display, Cool Edit Pro displays the spectral information in the Analysis window. When you view stereo data, the left channel is shown in cyan, and the right in magenta.

Statistics (Figure 14.7)

Analyse/statistics produces information about the whole waveform or that section that is selected.

Use the Statistics dialog to get the following details about the current waveform:

Minimum/Maximum sample value

Minimum, in this context, does not represent the 'quietest sample'. It couldn't, as every cycle of audio goes through zero. What it does represent is the peak negative-going excursion of the waveform. In both cases, clicking the arrow will take you to where this maximum or minimum is.

Peak amplitude

This combines negative and positive and expresses the peak level in decibels. A value of 0 db indicates a fully modulated wave file. Negative values indicate how much the level would be increased if normalized. Clicking the arrow will take you to the point the peak amplitude occurs.

Possibly clipped samples

A sample which has the value of the highest possible, or lowest possible, value can indicate full modulation.

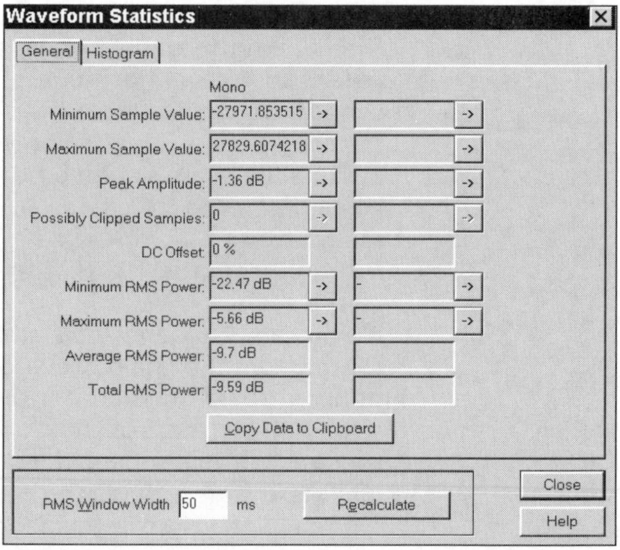

Figure 14.7

But, because no higher number can be represented, it may indicate that clipping – digital overload – has taken place. Clicking the arrow allows you to examine the waveform to see if there really is a problem by taking you to the point.

DC offset
This reports the DC offset to within 0.001 per cent and represents a test of how well the manufacturer has set up your sound card. Each sound card input may have a different value. Ideally they should all be zero.

Minimum/Maximum/Average/Total RMS power
This is roughly equivalent to the area of the waveform display. A very peaky waveform may consistently touch 0 dB but it may have less power than a continuous sound of lower level. The arrows will take you to minimum and maximum points. These may well be in different places from the minimum and maximum amplitudes.

Histogram tab (Figure 14.8)
This is another way of seeing the distribution of frequencies in your recording.

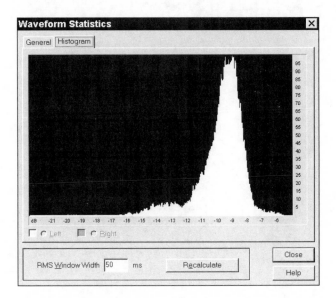

Figure 14.8

15

Archiving

15.1 Selection

Archiving often involves difficult decisions. If you keep everything, then you soon will be buried in old recordings. To be any use, they need to be catalogued. This is a lot of work.

Yet, so often, the interview that is thrown away is the one that turns out to have commercial value because of later fame or notoriety acquired by the person interviewed.

15.2 Compact disc

The ease with which computers can burn CDs has simplified the matter. A CD blank can contain about 80 minutes of audio as CD audio The discs themselves take up relatively little space, especially if kept in folders rather than conventional 'jewel cases'. Plastic sleeves punched for insertion into standard lever arch folders allow the discs to be kept with any paper documentation.

Keeping your material as CD audio means that it is easily accessed using ordinary compact disc players. The audio can be reloaded into the computer faster than real time ($\times 12$ or more). However, editing information is lost and there is a small loss of quality for each generation of copying. This is due to the compromise nature of CD audio recording, which was never designed as a data storage format.

The Track At Once option of much CD burning software has the advantage that it allows you to burn multi-session CDs. This could be useful for archiving, as you can add a track and remove the CD. On another day you replace the CD and add another track. While you are doing this the CD will be readable on your computer but NOT on an ordinary CD player. When you have finished adding tracks the CD is 'closed', or 'fixed'; no more tracks can be added but the disc now becomes readable on an ordinary CD player.

However, a better strategy for archiving is to allocate hard disk space for your archive material and then burn the CD in one go. Make two copies for real security. Ideally, archived material should be kept as computer files in CD-ROM form as this is error free.

15.3 CD-ROM

CD-ROM is designed as a data storage format at the expense of a little storage capacity; a little over 700 Mbytes of audio data for an 80 minute CD-R blank. For most medium-length items, this is enough to contain all the material used in the editing session, including the edit decision lists (session files in Cool Edit speak). This means that you can access the complete editor information and can 'unpick' edits or access the original uncut material. The data CD can also contain word processor and text files and even pictures associated with the item. This may include transcripts if they were prepared for editorial, copyright, or World Wide Web use.

On the other hand the convenience of having a CD audio version is very great, so a good compromise is to burn one of each. The audio CD becomes your every day 'library' copy. You can take it home, or to someone's office, and listen to it on any CD player. The CD-ROM is kept on the shelf and used only to transfer material to be reused. If the CD audio version becomes damaged then a new one can be made from the data files on the CD-ROM version. *In extremis*, a new data CD-ROM of the audio can also be made from the audio version.

15.4 DAT

DAT cassettes can be used for archiving. Audio cassettes have 2 hours of capacity at 44.1 kHz or 48 kHz. They have twice that at their half speed setting. This records at 32 kHz sampling rate. Using DATs manufactured for computer backup, you can get 4 hours or more onto a single cassette at full quality (8 hours or more at LP speed). However, actually using them for computer backup is a much better option.

DAT can be useful as a way of archiving 48 kHz masters in a way that can be played as audio on conventional machines. DAT has a number of disadvantages.

- The tape is very fragile and is prone to humidity problems. When a DAT gets mangled in a machine you have lost everything. While rescuing it is technically possible, this is a very expensive and time-consuming activity with no guarantee of success.
- While it may sound trivial, a substantial disadvantage is the small size of the cassette. There is little room to write an informative label on them and they have also been known to disappear down the backs of chairs! The CD with jewel case or folder has a definite advantage here.

15.5 Analogue

It is possible to archive material onto analogue media such as compact cassette or reel-to-reel. This has no advantage whatsoever except, perhaps, as a short-term compatibility with an existing system. Sooner or later, the existing system will have to be transferred to digital; it makes sense to do it sooner rather than later.

15.6 Computer backup

Just as all your word processor files and web transcripts can be transferred to the same CD-ROM as your audio, so this can be done using whatever computer backup system you have – you do have a computer backup system don't you?

All computers crash. All computer systems lose data. No one believes in backups until the day they lose a day's, a week's, a month's, a year's work. Companies have gone bankrupt because of poor, or no, backup strategy.

A comprehensive backup strategy will also take into account the possibility of theft and fire. Programs are (usually) replaceable. Data files represent investment of time and money. They cannot be replaced unless you have a backup.

Even if you do have a backup, do you know how to restore your files? A cynic would say that the world is full of good backup systems which are rubbish at restoring. Beware, while it is easy to protect against the theft of information by password protection and encryption, this protection is lost if the password is written on a Post-it® note stuck to the side of the monitor. Equally, if the computer is lost in a fire then so will the Post-it® note! You must have a safe place to keep the password.

Choosing a password

Passwords should be easy for you to remember but difficult for others to work out. If you must write a password down then leave it in a secure place. Avoid the obvious; 'secret' is one of the first things that a malicious hacker will try. They will also try the names of your spouse and any children.

For the best protection, the word should be at least six characters long and not be a dictionary word. It should include non-alphanumeric characters such as '&$* @%'.

The password may be case sensitive so that 'rad $38' is treated as a different password to 'Rad $38' (a capital 'R' is used in the second but not the first). The CAPS LOCK key being on can cause initial panic when the password you are so confident about is not accepted!

If you are an employee, then your company should have a system where your password(s) can be kept securely against the day when you meet your demise in a car accident (or, with long-term data, they may need to recover the data even if you left the company 2 years prior to the disaster).

There are a plethora of hardware data storage systems, some based on tape cartridges and some based on disk cartridges. Because of the possibility of theft or fire, the backup must be on some form of removable medium. Tape cartridge systems have the disadvantage of a relatively short cartridge life. The tape wears. Some recommend cartridges being replaced after anything from 20 to 100 uses (this includes DAT).

With the dramatic fall in price of hard disks it is worth considering using standard high capacity hard disk drives mounted in removable caddies for your backup.

Your backup regime should rotate round at least three removable media – the so-called grandfather, father, son strategy. Again, because of fire, you should store them away from your computer, in another room or even another building, especially with mature commercially valuable data. One way of achieving this is to backup, via a network connection, to a remote server. This can use a local area network or, if you have fast enough access, the Internet. For the amount of data involved in audio files you will need, at least, ISDN speeds and, even so, this will involve overnight backup runs. Talk to your system administrator about preferred ways of off-site backup.

For most purposes, taking the backup home with you overnight can be an effective way of meeting the backup safety criteria.

16

Tweaks

16.1 Mouse and Favorites

Computer programs always have options that allow you to tweak how they work so that you can make them easier to use. Cool Edit Pro is no exception.

Your choice of mouse will make a difference to Cool Edit Pro as it exploits a wheel mouse so that the wheel can be used to zoom in and out in the single wave view and to move up and down the track of the multitrack view.

The Favorites menu can be built to make a collection of your most commonly used actions. Use Edit Favorites to create, delete, edit, and organize items appearing in the Favorites menu.

Edit Favorites can instantly call up any customized Cool Edit Pro Transform or Generate effect, Script, or even many third-party tools. The menu can also contain sub-menus for easy organization.

16.2 Settings (Figure 16.1)

This tabbed dialog is the key to customization of Cool Edit Pro. This varies from what colours are used by the display to the arcane settings of MIDI and SMPTE parameters. It is here that the output of the single wave view editor can be selected to different sound card outputs if required.

16.3 Help (Figure 16.2)

This book is not intended to be a Cool Edit Pro manual. Syntrillium have already written that. Not

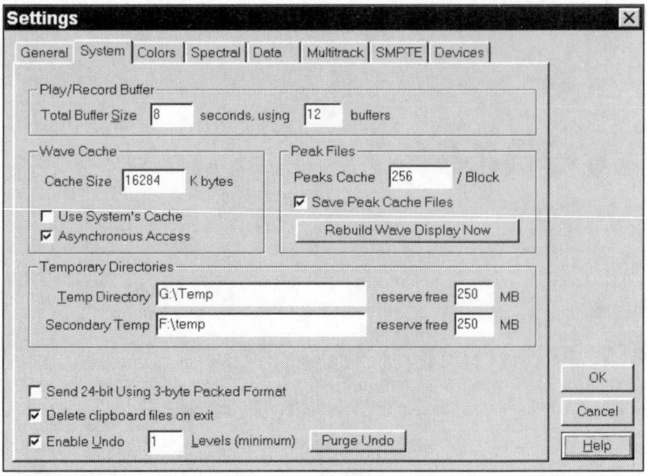

Figure 16.1

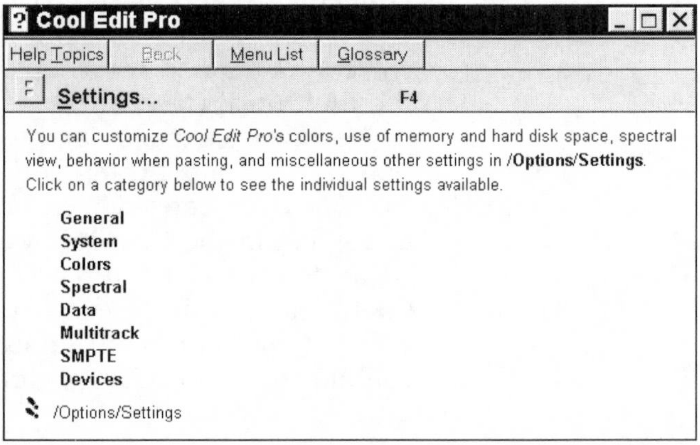

Figure 16.2

only is it on the CD but is also available in printed form on request.

The program also comes with substantial help files which will hold your hand through the various options. Illustrated is the help topic for the Settings dialogue. Each tab heading is listed. Clicking on a heading will take you to a detailed description of the options in that tab. That too will have clickable headings for each option. Right clicking on a topic allows you to print it or, more usefully, copy it to the clipboard to be added to a text file. In this way you can make a file of information on the areas that worry you, or you can even write your own manual!

16.4 Icons

Some people love icons, others stare at them uncomprehending. If you like icons then you can choose which groups of icons for shortcuts will be shown at the top of the screen. At the very least, you will probably want to keep the file group as this includes the icon to toggle between single wave view and multitrack. If you hate icons then you can remove them altogether. Normally the easiest way to tweak the icons is right clicking on the toolbar but, if you have no icons selected, then this is not possible and you need to use the Options/toolbars menu item.

16.5 Keyboard shortcuts (Figure 16.3)

Cool Edit Pro comes with some keyboard shortcuts already defined. You can change these or add more with the 'Options/Shortcuts (keyboard and MIDI triggers)'. Set up the actions you use most as simple key press combinations.

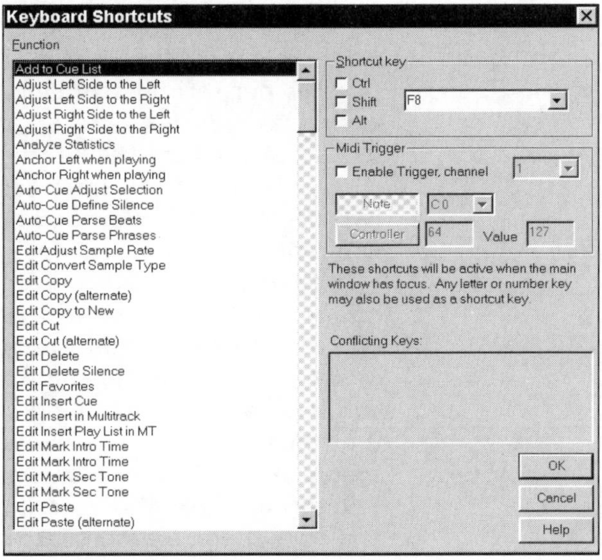

Figure 16.3

Appendix

Clicks and clocks

When you copy digitally to your computer, are you plagued with clicks and plops? Is there a regular beating sound where there should be digital silence? If so, you are probably suffering from unsynchronized word clocks. Arcane though it sounds, the problem is simply solved.

Copying digitally from one device to another is, in many ways, easier than using an analogue link. There is no level adjustment to be made and quality is assured, provided the link is made by good quality cable.

Yet, there is an extra factor that it is vital to get right; this is what is known as word clock synchronization. You will be using 32 000, 44 100 or 48 000 samples per second sampling. These figures are also known as the word clock rate. 48 000 samples per second also means 48 000 digital words per second. For a 16-bit stereo signal each word will consist of two 16-bit samples (left and right) packaged together with additional 'housekeeping' data.

The word clock synchronizes the whole system; it can be thought of as the conductor of an orchestra. When you are copying from digital recorder to digital recorder, there are no complications. It is only when you have a device handling several digital signals at the same time – like a digital mixer or a computer sound card – that problems arise.

Returning to the conductor of an orchestra analogy, there are some pieces of music – Mahler symphonies, etc. – that feature offstage bands that cannot see the conductor in the hall. Unless something is organized, they will not be able to play in time with the hall orchestra. A common modern solution is to have a closed circuit TV connection, so that a second conductor can synchronize his beat to the image of the conductor in the hall.

So it is with digital systems. If your card can handle other inputs at the same time as the digital one then you need to be concerned with word clocks. The 'conductor's beat', aka the word clock, is sent as part of the digital signal; so we have our 'closed circuit TV'. Normally the sound card will generate its own 'beat'. This you set when you choose the sampling rate. While this is set to match the rate coming in from the external playback, there is no synchronization between the two. Even with crystal control, the chances of both clocks being precisely the same frequency are remote, so every few seconds they drift one word apart. The error resulting from this is what causes the clicks or plops. The answer is to tell the sound card to 'look' at the word clock incoming from the DAT, or Minidisc machine, feeding it. The sound card management software should have this as an option, alongside the sampling rate selection.

Figures A.1 and A.2 show drop-down menus for a multitrack card which can handle either optical or electrical digital inputs (the selection between the two is made in a separate menu).

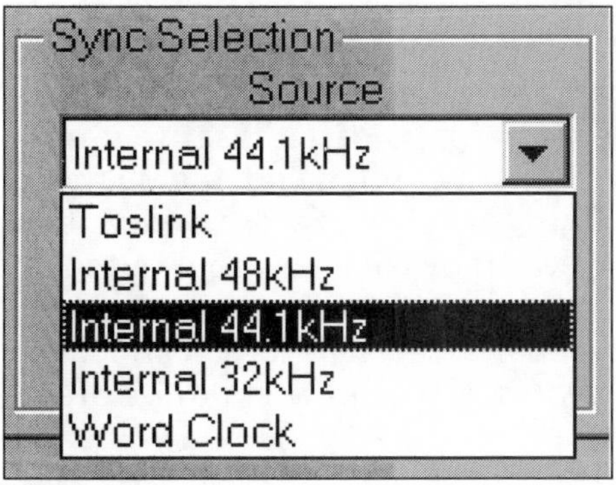

Figure A.1

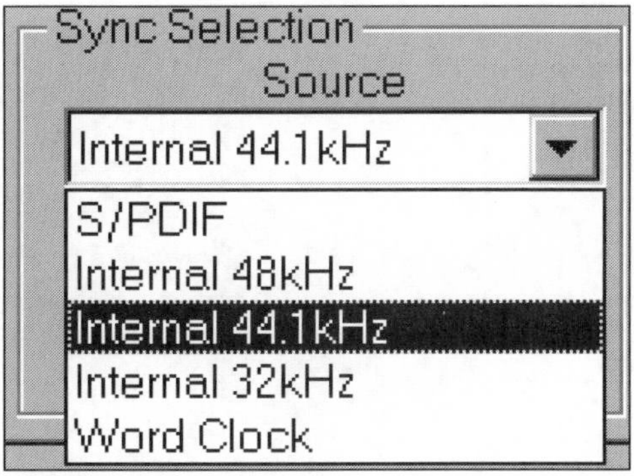

Figure A.2

If you select the S/PDIF, or TOSLINK, option then the sound card is controlled entirely from the external bit stream; it will follow the sampling rate as part of the synchronization.

Beware, that you have to switch synchronization back to internal once you stop using the external machine. As soon as it is disconnected, you will have no clock and the sound card may either not produce any sound, or sound at the wrong rate and pitch.

Another trap is that an external machine may default to a different clock, when it is not playing back. This means that, having very successfully copied a 44.1 kHz recording, the clock may go to 48 kHz when you take the DAT recording out. If you have not reset your sound card back to an internal clock setting, then the files played back through your editor may well play at the wrong speed and pitch. In this case faster and higher.

Word clock in/out

The illustrated sound card has another option which is specifically labelled word clock. This is for a fully professional installation where a separate high quality word clock generator is used. This separately synchronizes a number of separate digital devices, essential when dealing with a digital mixer.

Equally, when you have more than one digital sound card, you have to synchronize them by connecting the word clock output of the card you have designated as the master to the word clock in of the second card. The second card will always synchronize to the master provided it is set to use the word clock input, including following the sampling rate settings. If a third, or fourth, card is added, then the word clock can be 'daisy chained' from output to input of the next. Word clock connections are usually made with professional video-style BNC connectors. They keep the cards in synch even when there are no audio data present.

Glossary

AB Stereo Often used to distinguish MS stereo from the convention signal using left and right signals. In the context of microphones it often implies the use of spaced omnidirectional mics rather than a coincident pair.

ADAT A digital multitrack recording system that gives 8 tracks on an S-VHS videocassette.

ADPCM (Adaptive Differential Pulse Code Modulation) Conventional Pulse Code Modulation stores the values of a waveform as a series of absolute values. Differential PCM does not do this but, instead, sends the data as a series of numbers indicating the difference between successive samples.

AES (Audio Engineering Society) Among other things, the AES lays down technical standards. In the context of this book they are best known for a professional standard for conveying digital audio from machine to machine which has also been adopted by the EBU.

AES/EBU A professional digital audio standard for transferring digital audio between machines. This is balanced and uses XLR connectors. The data format is similar but not identical to S/PDIF which can see an incoming AES/EBU format signal as copy prohibited.

AIFF Apple AIFF (.AIF, .SND) is Apple's standard wave file format and is a good choice for PC/Mac cross-platform compatibility.

ALC (Automatic Level Control) See AVC.

Aliasing Spurious extra frequencies generated as a result of the original audio beating with a frequency generated within the audio processing system (usually the sampling frequency in a digital system). Filters are used on the input to prevent this but these filters themselves can produce degradation of the signal unless very well designed.

Analogue Audio Until digital techniques came along audio was conveyed and recorded by using a property that changes 'analogously' to the sound pressure. This property might be electrical voltage, magnetization or how a groove wiggled. Digital audio replaces this by a series of numbers.

AV (Audio Visual) AV standard hard drives are able to cope with long runs of data, such as a long continuous audio recording, without stopping to recalibrate themselves for temperature variations.

AVC (Automatic Volume Control) Often found as an option on portable recorders. This automatically adjusts the recording level from second to second. This can cause trouble when editing as the background noise will be going up and down. However, modern AVCs work surprisingly well. The background matching becomes a trivial problem when editing with a PC digital audio editor.

Balanced Normal domestic audio connections are unbalanced; a single wire carries the audio which is surrounded by a screening braid connected to earth as the return circuit. These circuits are prone to pickup of unwanted signals as well as high frequency loss when used beyond about 5 metres. Balanced circuits use two wires to carry the audio (still within a screening braid). The audio in the wires is going in opposite directions – as one wire goes positive the other goes negative. Interfering signals induce in the same direction on both wires. The input circuit is designed only to be sensitive to the difference in voltage between the two wires and therefore ignore the induced interference.

Barrier Mic Barrier mics are designed to be placed on large flat surfaces rather than suspended in free air. These are often referred to as PZMs (Pressure Zone Microphones) after a commercial version.

bel See Decibel.

Binaural Current practice is to use this term to mean two channel audio balances intended to be heard on headphones.

BIT (BInary digiT) Digital/PCM systems use pulses which indicate either an 'ON' or 'OFF' state. Each individual piece of data is known as a bit.

BNC A professional video/digital audio connector with a locking collar.

Byte A group of 8 bits (allegedly a contraction of 'By Eight'). This is the standard measure of the capacity of digital systems.

Capacitor An electrical component that can store electrical charge, formerly known as a condenser. They consist of two parallel 'plates' separated by an insulator. The plates are so close together that when they are charged the positive charge on one plate is attracted to the negative charge on the other. The closer they are together then the greater the attraction. This increases the amount of charge that the device can store. A practical capacitor's plates are in fact metal foil sheets separated by a sheet of thin insulator rolled into a cylinder, rather like a Swiss roll. This reduces their size and gives them a cylindrical appearance. A specially constructed capacitor forms the basis of electrostatic microphones.

Cardioid (Heart shaped) The most common of microphone directivity shapes. (See page 39.)

CD (Compact Disc) When used without qualification this is taken to mean a standard audio CD. Subsequently adopted for data use in computers, this has led to many variants (see below).

CD-R (Compact Disc Recordable) is a recordable CD which cannot be erased.

CD-ROM (Compact Disc Read Only Memory) is a CD containing data rather than audio.

CD-RW (Compact Disc Read–Write) is a recordable CD that can be erased. While they can be recorded in audio format, most domestic CD players cannot play them.

Chequerboarding A technique for mixing sections of audio or video. (See page 101.)

Cinch plug Another name for phono plug.

Clock All digital systems have a reference clock which acts rather like the conductor of an orchestra to keep everything in sync. (See Appendix.)

Clone In the digital audio context this is used to mean making an exact sample for sample copy. This is not possible with systems like Minidisc, which use lossy compression systems.

Coax, Coaxial plug A generic term for a connection that uses a cable where one or more conductors are surrounded by a wire braid that helps screen out interference. In the UK, this term is most often used for the plug used for television aerials. In the audio context, some companies use this term for a phono plug.

Coincident pair A stereo microphone technique using directional mics placed as close together as possible.

Compact Cassette The proper name for the ordinary analogue audio cassette; undoubtedly the most successful audio recording medium ever invented.

Compression: Audio Compressor limiters are the most used effects devices in the studio. They can be thought of as 'electronic faders' which are controlled by the level of the audio at their input. (See page 130.)

Compression: Data There are two types of data compression; Non-Lossy and Lossy.
Non-Lossy: The first, and traditional, form of compression reduces the data to be stored on a disk or sent via a modem. The best known format is the Zip format. A Zip file can be uncompressed to recreate the original data without any change or error.
Lossy: Graphics and audio are often compressed using formats that approximate the data using assumptions about how we see and hear. (See page 181.)

Condenser Microphone Condenser is an old term for capacitor.

Cool Edit Pro A commercial audio editing package for Windows, combining both linear and non-linear editing.

DAC (Digital to analogue converter)

DAO (Disc At Once) A technique of burning a CD in one go. This has a number of technical advantages, notably giving the ability to control inter-track gaps or even recording audio into those gaps.

DAT (Digital Audio Tape) Originally a generic term for 'Digital Audio Tape Recorder'. This is now used specifically for a format developed by Sony, supported by scores of other manufacturers, that has become popular amongst professionals and semi-professionals alike for mastering digital audio.

DATA Plural of 'datum'. From Latin 'things given'. The word is often erroneously used as a singular, i.e. 'The data are' not 'The data is'.

DAW (Digital Audio Workstation) A dedicated computer audio editor with specialized controls and software.

dB Decibel

dBA Decibels are used widely in the field of acoustics, audio and video. Subtly different scales are used and the different types of decibel are indicated by a suffix. The dBA is used in the field of acoustic measurement.

dbx A commercial company that is best known for developing a popular analogue noise reduction system.

DC Offset Poor analogue to digital converters can have a DC offset that 'pushes' the audio away from being centred around zero volts towards either the positive or negative. This is a major cause of clicks on edits. (See Chapter 6.3, Figure 6.4.)

Decibel One-tenth of a bel. The normal way of measuring audio. It is a logarithmic system using a standard reference level. Values are expressed as a ratio of a standard level. The bel itself is too large a unit to be convenient for audio. The Richter units used for measuring earthquakes are identical to bels but with a rather louder reference level! (See page 7.)

Delta Modulation A technique where the difference between samples is sent instead of the absolute values usually sent by PCM.

Digital Audio Sound pressure level variations are represented by a stream of numbers represented by pulses.

Digital Audio Workstation A dedicated computer audio editor with specialized controls and software.

Digital Versatile Disc (DVD) This is the likely successor to the compact disc. It uses similar technology but takes advantage of technical developments since the CD was introduced. Recordings are made at a much higher density, eight times greater than CD. Additionally the DVD can not only be double-sided but also each side can be made up of two layers. This gives massive data capacity, enough for full length films with 5.1 channel audio. The '.1' is a, not very good, engineering joke. There are, in fact, six channels of audio but the sixth channel is a low bandwidth one used for low frequency effects.

DIN (Deutsche Industrie-Norm) German industrial standard. This includes the audio/MIDI/computer DIN plug. A standard sized case and connector containing a number of pins. The most common type met in the audio context is the 5-pin 180°.

Disc, Disk A convention has grown up where disc-based media using a magnetic medium are spelt as 'disk' (with a 'k'). Optical recordings like CD and Minidisc, as well as gramophone records, remain spelt as 'disc' (with a 'c').

Dither A low level signal, usually random noise, which is added to the analogue signal before conversion to digital. Its effect is to reduce the distortion caused by quantization.

Dolby Dr Ray Milton Dolby – possibly the most influential individual in audio. His company, Dolby Laboratories, began by making audio noise reduction systems. The first, a professional system, became known as Dolby A-type. A simplified system, called Dolby B-type, revolutionized the compact cassette medium for consumers. Dolby C-type gives about 20 dB noise reduction compared with the B-type system's 10 dB, albeit at the cost of poorer compatibility when played on machines without decoding. Dolby SR (Spectral Recording) is an enhanced professional system that can give better than 16-bit digital performance from analogue tape. Dolby S is a powerful compact cassette system based on a simplified version of SR. As other companies have to obtain a licence to use the Dolby trademark, the company has become the effective setter of standards for cassette machines, as minimum audio performance is set by Dolby Laboratories to allow their systems to be used.

Dolby Digital Started in the cinema industry, this is the digital multi-channel sound format most widely used on DVD and with digital television.

Dolby Stereo Dolby Laboratories devised a system for putting stereo onto optical release prints in the cinema. This was combined with a phase coding system that allows surround information to be heard. The Left total and Right total channels can be decoded to give Left, Centre, Right and Rear loudspeaker information. In the early 1990s the term 'Stereo' was dropped, and the analogue film format is now simply referred to as 'Dolby'.

Dolby Surround The domestic version of Dolby Stereo as found on video cassettes, CDs, TV broadcasts and video games.

Drop In Switching from play to record while running to make an electronic edit.

Drop Out 1. A momentary loss of sensitivity in an analogue recording medium. Digital systems can correct or conceal errors resulting from drop outs in the medium. However, if they fail you will hear a mute instead.
2. Switching from record to playback to end a drop in.

DTMF (Dual Tone Multi-Frequency) These are the tones that modern phones make when dialing. They are used to code the digits 0–9 as well as the special system codes of '*' and '#'. Four extra codes are also available known as 'A', 'B', 'C' and 'D'. The sixteen possible combinations are achieved by sending two frequencies (hence 'dual tone') out of a possible selection of a total of eight.

DVD (Digital Versatile Disc)

Dynamic Microphone An alternative word for a moving-coil microphone.

Dynamic Range In audio systems the dynamic range available is determined by the number of bits used to measure each sample. In theory for every bit extra another 6 dB of signal to noise ratio is gained.

EBU (European Broadcasting Union) A trade association for European broadcasters, which also sets technical standards.

Echo Although this is often used interchangeably with the term reverberation (or 'reverb'), they are technically different. Echo is where you can distinguish individual reflections (echo ... echo ... echo) while reverberation is where there are so many reflections that they merge into one continuous sound. (See page 134.)

Edit Decision List (EDL) In a non-linear editing process where the audio files are not altered. Instead a list of instructions (the Edit Decision List) as to what section to play when, at what level, etc. is created which causes the hard disk to skip around and produce apparently continuous audio.

EDL (Edit Decision List)

EIDE (Enhanced Integrated Drive Electronics) The most widely used (and hence cheapest) form of hard drive. Most PC motherboards are equipped to handle four drives. The usual alternative is SCSI.

Electret A form of capacitor which remains charged permanently. This means that when constructed as a microphone, it does not need a polarizing voltage. Typically a simple 1.5V AA battery is used to power the built-in amplifier.

Electromagnetic Devices where magnetism is used to create electricity or electricity is used to create magnetism. Within audio, a moving coil (dynamic)

mic is used to create electricity (the audio signal) by the diaphragm pushing and pulling a coil of wire between the poles of a magnet.

Electrostatic Electrostatic microphones use a diaphragm that is one of the plates of a capacitor held charged by a polarizing voltage. (See page 41.)

EQ Pronounced 'Eee' 'Cue'. A widely used abbreviation for 'equalizer', a term for devices like tone controls which modify the frequency response of an audio system. Such devices were originally used by engineers actually to equalize, or correct audio deficiencies in land lines and recording systems. They were then borrowed by studio operators to improve their recording mix. Nowadays equalizers are designed specifically for studio use.

Error Concealment A technique where errors can be detected but not corrected. Instead they are concealed often by replacing the sample with an average of the samples either side. This is called interpolation (see below).

Error Correction As digital audio is a series of numbers, extra numbers can be added having been generated by various mathematical means. At the receiving end these numbers can be generated again from the incoming data. If they are different from the extra numbers sent, then error has been detected. With suitable maths, the errors can often be corrected. Because the binary nature of the signal represents only '0' or '1', it is clear that, if the incoming value is established as being wrong, then the correct value must be the only other value.

Interpolation: is a technique where when an error is detected a value between the preceding and following value is substituted, whereas full error correction actually reconstitutes the data.

Interleaving: Errors can be made easier to correct by interleaving the data so that they are physically spread out on the medium so that a single drop out in the medium does not produce a single burst of errors.

fff 'f' is a music term for loud (Italian: *forte*); the degree of loudness is indicated by the number of 'f's, thus fff is very loud.

FFT (Fast Fourier Transform) A mathematical way of defining a filter.

Figure of Eight Term used to describe a microphone that is sensitive front and back but dead at the sides.

Flanging Phasing with continuously varying delay.

Flutter Rapid variation of pitch, often caused by a dirty or damaged capstan pulley on an analogue tape machine. A digital recording will be free from this fault.

Flutter Echo Term used by those of us who tend to use the terms 'echo' and 'reverb' interchangeably to show that we are really talking about echo.

FM (Frequency Modulation) A way of sending data, or audio, using a carrier frequency which is varied in pitch. This is used by FM Radio, VHS video cassette, Hi-Fi sound, many hard disks, etc.

FX A widely used abbreviation for 'effects'.

Gain Another word for amplification.

GIGA 1000 million, hence 1 GigaHertz = 1 000 000 000 Hertz = 1GHz.

Glitch A discontinuity in sound, due to data errors.

Gun Mic A generic term for very directional mics using a long tube and phase cancellation techniques. Also known as shotgun or rifle mics. Because they are very sensitive to wind noise they are usually concealed in long furry sausage-like windshields.

Hyper-cardioid A microphone which has a slightly narrower front pick up compared with a cardioid. The penalty is that there is a reduced sensitivity lobe at the back, giving a dead angle a few degrees to the side of this.

IDE (Integrated Drive Electronics) A way of connecting hard drives to personal computers. It is very cheap and has become very popular and hence become even cheaper. Although the technology has much improved it is regarded, by many, as inferior to SCSI as it uses the computer's processor and slows down programs that are being run at the time. It has been said that IDE works in a way analogous to arriving at a shop counter and asking for an item and saying 'I'll wait' whereas SCSI allows you to go away and do something else because it will 'deliver'.

Image The perceived location of a single source within the sound stage. The image may be narrow (panned mono) or wide (string section). It may be precise or blurred.

Interleaving Error correction technique (see under Error correction above).

Interpolation Error concealment technique (see under Error correction above).

ips Inches per second.

ISDN (Integrated Services Digital Network) Sometimes ironically referred to as the 'It Sometimes Doesn't Network', this gives you a direct digital connection to the phone network. There are various audio coders that allow good quality down, what is effectively, a telephone line. Different organizations have standardized on different systems and some manufacturer's gear is not fully compatible.

Kilo (K) The word Kilo when used in the context of digital systems usually means 1024 *not* 1000. This represents the maximum value of a ten-bit word, i.e. 2 to the power of 10. This is not cussedness, but is used because it represents a very convenient unit. So the term 64K will mean 65 536. Some publications use the abbreviation 'K' to mean 1024 and 'k' to mean 1000.

Limiting Compression of greater than 10:1.

Line In An input to be fed by amplified audio rather than a microphone.

Line Level Domestic outputs tend to be −8 dB with professional outputs being +4 dB. 0 dB is 0.775 Volts. This strange value dates from telephone technology, where 0.775 v gave 1 milliwatt into a 600 ohm circuit.

Line Out Amplified output of a device.

Linear Editor An editor where the audio files themselves are altered by the editing process.

Lossy/Non-Lossy See Compression: Data.

LP (Long Play) Usually used to refer to 12″ gramophone records. However video cassettes, DAT and Minidiscs have a long play mode.

Mastering The general term for transferring studio material to a final stereo 'master' which will be used to generate the copies sold to the public.

Mega 1 million, hence 1 megaHertz = 1 000 000 Hertz = 1Mhz (note upper case 'M'; lower case 'm' means milli = 1/1000th when used as a prefix).

Mic Preferred British abbreviation of the word microphone.

Micro One millionth, hence 1 microsecond (1 μS). Used colloquially to mean small computer.

MIDI (Musical Instrument Digital Interface) A standard system for communicating performance information between synthesizers and computers.

Milli One thousandth, hence 1 millimetre = 1 thousandth of a metre (1mm).

Monaural Sometimes used to mean monophonic but really means listening with one ear!

Monitoring speaker Monitoring speakers are designed to be analytical and reveal blemishes so that they can be corrected. As a generalization, hi-fi loudspeakers make the best of what is available.

Mono Contraction of monophonic; also contraction of monochrome (i.e. black and white pictures).

Monophonic Usually contracted to 'mono', this is conventionally derived from the stereo signal by a simple mix of left and right channels. Beware of some portable tape machines which have a switch labelled mono. This often means that only one input is fed to both legs of the recording *not* that the two inputs are mixed and fed to both legs. It is very easy to end up with an interview tape that only has questions on it if a reporter is not aware of this.

Moving Coil The most common type of microphone and loudspeaker. A coil of wire attached to a diaphragm is suspended in a magnetic field. If the coil is moved by pressure on the diaphragm a small voltage is generated. If a current is passed through the coil, the diaphragm will be moved.

MPEG (Motion Picture Experts Group) MPEG is a series of lossy compression systems for audio and video.

MS (Middle-Side) A stereo technique where instead of the left and right information being used for the two audio channels, the middle and side are used. The middle signal corresponds to the mono signal. The side signal corresponds to the amount of 'stereo-ness'. It is zero for centre signals and at a maximum for sounds coming from the extreme left or right.

Multiplex Generally any method of carrying several signals (e.g. stereo left and right) on a common circuit. Digital radio and television are broadcast using multiplexes that carry a number of channels; how many depends on the quality required.

Nano 1 thousand millionth, hence one nanosecond (1nS).

NICAM (Near Instantaneously Companded Audio Multiplex) Digital system used in the UK to add stereo sound to television broadcasts.

Noise reduction Analogue recordings are often made using noise reductions systems like Dolby and dbx. They boost the signal on record and apply a correction to this on playback. In the process noise and hiss are correspondingly reduced. In the digital editor context noise reduction usually refers to various software solutions that remove clicks, hiss or noise from an existing recording – usually a transfer from an analogue original.

Non-Linear Editor An editor which does not alter the original audio files. Instead it uses some form of edit decision list to instruct the computer to jump around the hard disk reproducing and mixing audio as required.

NTSC (National Television System Committee) American colour television system.

Nyquist Limit This is named after Harry Nyquist the Bell Telephone Laboratories' theoretician who first enunciated the principle that you have to sample at a frequency at least twice that of the highest you intend to transmit.

Omnidirectional Omnidirectional microphones are equally sensitive to audio from any direction.

Out-of-Phase If loudspeakers are out-of-phase, this means that their diaphragms, instead of moving in and out together, move in opposite directions. This has the effect of blurring the sound image and reducing the bass response of a system.

PAL (Phase Alternate Line) Colour television system used by most countries in Europe. This is often used to imply a 25 fps frame rate, although 30 fps versions do exist. Many videocassette machines will output a PAL signal at 30 fps when playing an NTSC tape.

PC Originally a generic term for 'personal computer' this has been hijacked to mean a computer based on the original IBM design. While sometimes used to imply using a Microsoft operating system such as Windows, as opposed to Apple Mac, they can also be used for other operating systems such as BeOS, Unix, Linux, etc.

PCM (Pulse Code Modulation) The technique of sending numbers as a series of pulses.

Phantom volts Electrostatic (capacitor or condenser) mics need to be powered to work. Balanced studio mics are often powered by adding the DC voltage (usually 48 v) to the audio wires. Circuitry at each end separates the audio. Although the circuit behaves as if there is an extra wire for the power, it has no physical existence, hence the term phantom.

Phase A measure of the relative delay between two waveforms at the 1 cycle level. The positive-going zero crossing is described as 0° and the negative-going zero crossing as 180°. Where the difference between the two waveforms is exactly reverse – positive in one is matched by negative in the other – they are said to be 180° out-of-phase, or just out-of-phase.

Phasing This caused by selective cancellation of some frequencies either using a comb filter or, more often, by mixing two nominally identical signals with a short delay between them. In the late 1960s this was a much loved 'psychedelic' sound. If the delay is continuously varied this is called flanging.

Phono plug The RCA phono plug was originally designed to connect phonograph (gramophone) turntables to amplifiers. This has become a universal way of connecting unbalanced audio (and, often, video) largely because of the cheapness of the connector.

POTS (Plain Old Telephone System) A general term for ordinary phone lines which have limited capabilities compared with specialized systems using newer technology such as ISDN.

ppp 'p' is a music term for quiet (Italian: *piano*); the degree of quietness is indicated by the number of 'p's, thus ppp is very quiet.

Pre-emphasis A technique of boosting high frequencies on record or transmission; they are then restored on playback or reception. In the process any added hiss has top cut applied to it.

Program, Programme In this book 'program' is used in the context of computers. The British spelling is used in the context of Radio and TV programmes.

PZM (Pressure Zone Mic) See Barrier mics.

Quantizing The process of turning an analogue signal, like audio, into numbers.

Quantizing Interval The difference in voltage between quantizing levels.

Quantizing Levels The number of possible values into which an analogue signal may be divided or quantized.

RCA plug The RCA company originally devised a plug for connecting phonographs (gramophone turntables) to amplifiers. These plugs became known as phonograph plugs, abbreviated to phono plugs.

Reverb Contraction of 'reverberation'.

Reverberation The sound made by reflections from a room, or an electrical simulation of this. It differs from echo in that there are so many reflections that individual reflections are not discernible. Reverberation time is measured as the time taken for the reflection level to decrease by 60 dB (RT_{60}).

RIAA (Radio Industries Association of America) A trade association that also sets standards. Most often met is the RIAA standard for equalization of long playing gramophone records. When the disc is cut the high frequencies are boosted and the bass frequencies cut. This is corrected by reversing this on playback.

Ribbon Mic A microphone using a corrugated aluminium ribbon as a diaphragm. This is placed within the field of a powerful magnet. Once the standard microphone used by the BBC and other broadcasters, it is capable of extremely good quality especially on strings. It is advisable to take off your wristwatch before handling one as the strong magnetic field may stop it.

S/PDIF (Sony/Philips Digital Interface) Standardized digital audio connection format using an unbalanced audio connection and phono plugs.

SCMS (Serial Copy Management System) A specification for copy protection flags contained within domestic digital audio connections. While the system allows for no copy inhibition or full copy protection, its default is usually to allow only one digital copy; a SCMS equipped digital recorder will copy a digital recording set to allow one copy but the copy it makes will be set to no copying allowed. In practice the system is merely an annoyance for serious recordists working with original material. It can easily be defeated by analogue copying, copying via a computer, or via an interface box that allows the resetting of the SCMS flags.

Scrub editing A term now used to describe the traditional way of finding edits on reel to reel tape which involved 'scrubbing' the tape back and forth. Many users familiar with tape editing like to have this facility on a digital editor, although this usually soon turns out to be a security blanket.

SCSI (Small Computer Systems Interface) Usually pronounced 'Scuzzy'. A way of connecting personal computers. Originally used on Apple Macintosh computers, it is regarded as having many technical advantages over its rival, IDE, universally used by PCs. However, there is no reason why a PC cannot use SCSI (some PC mother boards have it built in, others can have a SCSI card fitted). Many people regard the extra expense worth the extra reliability it gives to audio recording. Both systems can be used within the same machine. The SCSI interface is also used for other devices like scanners. Seven devices plus the controller can be accommodated by a simple SCSI system, with fifteen devices available on more sophisticated systems.

SECAM (System En Couleur Avec Memoire) French Colour Television system. SECAM VHS tapes will play in monochrome on a PAL video cassette machine.

Serial A term used when data is sent down one wire, one bit at a time.

SMPTE (Society of Motion Picture and Television Engineers) In conversation, often referred to as 'Simty'; this body sets standards.

SMPTE/EBU The combined American and European time code standard.

Stereo, Stereophony Literally 'Solid Sound' from the Greek. While there is argument over a precise definition it seems generally to mean using two channels to give a directional effect. Good stereo will do more than this, giving a sense of depth as well as direction.

Table of Contents Compact Disc and Minidisc use a

table of contents to tell them where each track and index begins and ends.

TAO (Track At Once) Where a CD is burned one track at a time with one wave file per track. The laser is turned off between tracks.

Time code A code contained within a recording that identifies each part of the tape uniquely in terms of time.

Tinnitus A distressing condition caused by hearing damage where noises are generated within the ear, often perceived by the victim to be at a very loud level.

TOSLINK (TOShiba LINK) Optical connector used by many domestic digital audio devices.

Transient A short-lived sound, typically the leading and trailing edges of a note.

USB (Universal Serial Bus) A standard way of connecting apparatus to a computer that is supported by both PCs and Apple Mac computers.

Variable Pattern Mic This contains two mic capsules that can be combined in different ways to give a range of directivities. (See page 40.)

Varispeed A control for varying the speed of an analogue recorder, usually on playback. This not only changes the rate but also the pitch. Digital editors usually have software that will allow change to rate or pitch independently of each other. Small changes of less than 10 per cent are usually very successful. Larger changes can suffer from glitches and artefacts. As a result you are often offered a choice of software methods (algorithms) which have different strengths and weaknesses.

VHS (Victor Home System) Presently the most popular home video tape format. In its NTSC version this has long play and extended play modes in addition to standard play. PAL systems only have a long play mode. The audio is recorded in two forms. There is a low quality recording on a linear track on the edge of the tape; this is usually mono but some machines provide stereo sometimes with Dolby noise reduction. Additionally many machines record in stereo using rotating heads within the video signal. The actual sound uses an FM carrier with a noise reduction using high frequency pre-emphasis and 3:1 compression.

WAV The .WAV format originated from Microsoft as a simple format for storing audio for games, etc. The format has been expanded for professional use adding text fields, etc. If an exported .WAV file begins with a click on another piece of software then it is possible that this software is not reading the file correctly. There is usually an option to export the files without the text information which can solve the problem. There is also an ADPCM version of this format giving 4:1 lossy compression. This is best regarded as an end-user format.

Wow Slow variation of pitch, traditionally caused by an off-centre gramophone record or a slipping capstan pulley on a tape machine. Digital audio systems are entirely free of this unless it is deliberately introduced as a special effect.

XLR Connector widely used in the professional audio field.

XY stereo In the context of microphones this is sometimes used to indicate the use of a coincident pair of microphones as opposed to spaced (AB).

Zip drive Proprietary form of removable hard disk cartridge.

Zip format Lossless data compression format much used to reduce the size of program and data files on disk and sent via modems. Unfortunately the assumptions that it makes do not apply to audio and files can end up larger after Zip compression than before.

Index

 **FocalPress**

http://www.focalpress.com

Visitourwebsitefor:

- ❑ ThelatestinformationonnewandforthcomingFocalPresstitles
- ❑ Technicalarticlesfromindustryexperts
- ❑ Specialoffers
- ❑ Ouremailnewsservice

JoinourFocalPressBookbuyers,Club

Asamember,youwillenjoythefollowingbenefits:

- ❑ Specialdiscountsonnewandbest-sellingtitles
- ❑ AdvanceinformationonforthcomingFocalPressbooks
- ❑ Aquarterlynewsletterhighlightingspecialoffers
- ❑ A30-dayguaranteeonpurchasedtitles

MembershipisFREE.Tojoin,supplyyourname,company,address,phone/faxnumbers andemailaddressto:

USA
ChristineDegon,ProductManager
Email:christine.degon@bhusa.com
Fax:+17819042620
Address:FocalPress,
225WildwoodAve,Woburn,
MA01801,USA

EuropeandrestofWorld
ElaineHill,PromotionsController
Email:elaine.hill@repp.co.uk
Fax:+44(0)1865314572
Address:FocalPress,LinacreHouse,
JordanHill,Oxford,
UK,OX28DP

Catalogue

ForinformationonallFocalPresstitles,wewillbehappytosendyouafreecopyofthe FocalPresscatalogue:

USA
Email:christine.degon@bhusa.com

EuropeandrestofWorld
Email:carol.burgess@repp.co.uk
Tel:+1(0)1865314693

Potentialauthors

Ifyouhaveanideaforabook,pleasegetintouch:

USA
TerriJadick,AssociateEditor
Email:terri.jadick@bhusa.com
Tel:+17819042646
Fax:+17819042640

EuropeandrestofWorld
BethHoward,EditorialAssistant
Email:beth.howard@repp.co.uk
Tel:+44(0)1865314365
Fax:+44(0)1865314572